Forest Tribology and Anthropology

Vinod M Mhaiske
Vinayak K Patil
Satish S Narkhede
College of Forestry
Dr. Balasaheb Sawant Konkan Agriculture University,
Dapoli, Dist. Ratnagiri, Maharashtra

SCIENTIFIC
PUBLISHERS

Published by:
Scientific Publishers
5A, New Pali Road, P.O. Box 91
Jodhpur 342 001, India
E-mail: info@scientificpub.com
Website: www.scientificpub.com

Print : 2019

ISBN: 978-81-7233-966-1
eISBN: 978-93-86102-08-9

Printed in India

PREFACE

Forest Tribology and Anthropology plays an important role in the curricula of forestry education at various levels. The basic knowledge on tribes including characteristics, family, distribution, dialect, kinship, marriage, dormitory, religion, culture, magic, anthropological aspects of tribe and the tribal development through various plans, schemes and programmes is much needed for the forest officials for the integral planning and development of the forest and tribals.

This book is more useful for the students of forestry who are going to opt for the conservation and development of natural resources as a career. This book is written in a simple form based on the teaching experience of the subject to the under and post graduate students of College of Forestry, Dr. Balasaheb Sawant Konkan Agriculture University, Dapoli, Ratnagiri District, Maharashtra State, and from the syllabus of Forest Tribology and Anthropology in B.Sc. Forestry Degree Programme.

An attempt has been made to present the subject matter in a lucid language to be understood easily by the students of forestry. The latest data relevant to Indian tribes is incorporated to update the students' knowledge.

This book is more useful for the under and post graduate students of Forestry, research scholars, general readers and candidates who are preparing for the competitive examinations of Union Public Service Commission (UPSC), Indian Forest Service (IFS), Agriculture Research Service (ARS), Maharashtra Public Service Commission (MPSC), Range Forest Officer Examination, Bank Recruitment Examination, Railway Board, other State Administrative Services, Forester and Forest Guard examinations, etc.

We hope that readers preparing for competitive examinations and those in the education of natural resources field find it useful for academic as well as competitive examinations.

Authors

CONTENTS

1

TRIBAL: MEANING, CHARACTERISTICS, HISTORY AND RACE

1.1. INTRODUCTION

India is a land of varied cultures, religions and people. Unity in diversity is the *mantra* of the mother Earth. An impressive population of tribal people is found among all these diversifications. These tribal people are believed to be the true inhabitants of India. Almost all the States in India have significant presence of tribal people who constitute intense portion of the population of the State.

Since prehistoric times, India has been a country of multi-racial stock. Its different parts are occupied by various groups of people having their definite cultural traits and level of development. Among them, a number of groups are still in primitive stage and are away from the impact of modern civilization. Therefore, these people are called aboriginals. The popular names for tribals are *Vanyajati* (caste of forest), *Vanvasi* (inhabitant of forest), *Pahari* (hill dwellers), *Adimjati* (primitive people), *Adivasi* (indigenous people), *Anusuchit Janjati* (Scheduled Tribe (ST)) and so on. In the ancient literature of India, their names, in the ancient languages, are observed as *Nishadas, Sabaras, Bhills, Kollas, Kirtas, Dasas, Dasyus, Sudras, Dramidas* and *Dravidas*. In the Indian Constitution they are termed as 'Scheduled Tribe (ST)'.

1.2. DEFINITIONS OF TRIBE

The term 'tribe' is derived from the Latin word 'tribus'. Earlier, Romans used this term to designate the divisions in a society. Latter use suggests that it meant poor people. Tribals have been defined into various ways. Following are some of the notable definitions of tribes given by different workers:

- Tribe is a group speaking a common dialect and inhabiting a common territory – **W. S. Perry**
- Tribe is a small isolated, closely-knit society – **Crocker**
- Tribals have single government and act together for such common purposes as warfare – **Rivers**

- Each tribe possesses a distinctive culture that makes it off from the other tribes – **E. A. Hoebel**
- Tribe must have a specific name, unity of speech, specific territory and inbreeding nature – an endogamous habit – **Wissler**
- Tribal/Tribe is a collection of families, bearing a common name, speaking a common dialect, occupying or professing to occupy a common territory and is not usually endogamous, though originally it had been – **Imperial Gazetteer Definition**
- In the Constitution of India, tribal definition has been given in Art-342. The Constitution of India, Article 366 (25) defines Scheduled Tribes as "such tribes or tribal communities or part of or groups within such tribes or tribal communities as are deemed under Article 342 to the Scheduled Tribes (STs) for the purposes of this Constitution" – **Constitution of India**
- Tribe is a social group with territorial affiliations, endogamous, with no specialization of functions, ruled by tribal officers, hereditary or otherwise, united in language or dialect, recognizing social distance from other tribes, following tribal traditions, beliefs and customs, illiberal of naturalization of ideas from alien – **Majumdar**
- Tribe, as any collection of political local group, which occupies a common general territory, speaks a common language and possesses a common culture – **Gillin and Gillin**

On the basis of above given definitions, we can conclude that the tribe is a collection of families, bearing a common name, common living homeland, members of which speak the same language and observe certain taboos regarding marriage, profession or occupation, and have developed a well-assessed system of reciprocity and mutuality of obligation.

1.3. STRIKING FEATURES / PECULIAR CHARACTERISTICS OF INDIAN TRIBE

The important characteristics of tribe are definite territory, common language, blood relationship, endogamy, common culture, common religion, common name, political organization, taboos, totem, etc. The following are the major characteristics of Indian tribe:

1.3.1. Definite Territory

The members of a tribe permanently live in a definite territory. Common territory is an important feature of Indian tribe. They have territorial identity i.e., they occupy a common and well-defined territory, which serves as its homeland for tribal community. In the absence of these tribals, existence cannot be imagined e. g., an area/region occupied by *Gond* tribe is known as *Gondwana.*

1.3.2. Common Language

All the members of a tribe speak a common language or dialect. This common language not only binds them in strengthening unity, but also helps to increase strong community feeling among themselves. Most of the tribal languages do not have scripts. It is one of the most important sources of transfer of culture from one generation to the next generation. e.g., *Gond* tribe - *Gondi* language, *Korku* tribe - *Korku*, *Gavit* tribe - *Bhilli*, *Katkari* tribe - *Katkari*, *Kolam* tribe - *Kolami*, *Madia* - *Gondi*, *Warli* tribe - *Warli*, etc.

1.3.3. Blood Relationship

The members of a tribe are related to each other by blood. They claim to be originated from a common ancestor, which may be real or mythical. They are close blood relatives because of the endogamy of the tribes.

1.3.4. Endogamy

Another main feature of a tribe is its endogamous nature. The members of a tribe are required to marry within a group or within a specific tribe or similar social unit. The rule of endogamy permits a person to marry within his or her own group. The group may be a caste (in India), a class, a rule, an ethnic group or a religious group or even a village. The rule of endogamy stipulates that the partners must be from the same group. Thus, endogamy is a rule that prescribes marriage between a boy and a girl belonging to the same group.

Marriage outside the group is strictly prohibited. Tribes are usually endogamous group, though certain tribes have become exogamous after coming into contact with the external world. Within the tribe, there are sibs or clan, which may be exogamous. Endogamy helps the tribes to propagate their common culture from generation to generation.

1.3.5. Common Culture

All the members of a tribe are mainly guided by a common or homogeneous cultural pattern. This cultural similarity is an important factor to regulate the tribal organizations. Tribes do not adapt easily to change their culture because of the nature of their life. Common language and endogamy in tribes helps to promote common culture among tribal societies. Each tribal community is rich in their culture, folk tales, folk songs and folk stories.

1.3.6. Common Religion

All the members of a tribe have a belief in a common religion. They extend direct co-operation to each other, being inspired by the *mantras*: 'We belong to a particular tribe and our religion is one'.

1.3.7. Common Name

Every tribe has a common name by which it is identified. This name has great significance from sociological point of view, which binds the members of a tribe with a feeling of unity.

1.3.8. Common Occupation

Each tribe earns their livelihood from a specific source. Economic activities of the tribes differ from tribe to tribe. e.g., hill cultivation, cattle rearing, snake charming, hunting, plain agriculture, folk art, simple artisans, collection of Minor Forest Produce (MFP), acrobatics, handicraft, etc.

1.3.9. Political and Social Organization

Every tribe of India is known to be a strong political and social unit. These organizations are meant to maintain peace, harmony, unity and integrity in a tribe and to avoid notes of discord among its members and to protect them. The responsibility of the organization is vested upon an individual, known as *Mukhia.* He is the unique ruler of the tribe and is responsible for solving any dispute in a tribal society. *Mukhia* commands the tribe and his decision is final; the violation of this will be punished by tribal chief. All the members have fear, faith and devotion towards the *Mukhia.* The selection of this officer is either hereditary or through trial of strength. A political organization works for the following: (1) an establishment and maintenance of public peace; (2) the security and health or welfare of the body; (3) the doing of justice or equity; (4) the advocacy and defence of justice; (5) the recognition of the authority of law, supported as it was by the body of warriors; and (6) the use and preservation of the magic power.

1.3.10. Taboos, Totem and Myths

Tribes believe in taboos, which are certain forbidden things or restrictions communicated through verbal 'to do' or 'do not'. Tribals believe in totem, which are certain objects having mythical relationship with the members of the tribe. The totem is considered to be the originator of a particular tribe. It's killing and eating is a taboo for the people of the particular tribe. It is worshipped and respected everywhere. Marriages cannot be contracted among the male and female of the same totem, because they are thought to be having blood relationship. They also believe in certain myths, rudimentary types of religion and magic.

1.3.11. Others

According to Dr. Paul Chaudhari, some of the tribal characteristics are: Primitive and simple way of life, shy of stranger, distrust of a non-tribal, love for drink and dance, nomadic life, murder and crime is common, habitation in remote and inaccessible areas, hunting considered as a ritual, illiteracy, ignorance, no script for language, common occupation, etc.

1.4. HISTORY OF INDIAN TRIBE

Indian tribe is one of the earliest of the present inhabitants of the country. In the ancient literature, there has been a reference of *Janas* or communities such as *Savaras, Kallutas, Killos, Bhillas and Kinnouris*. Many tribal communities cited above may be the ancestors of the tribes. Tribals are indigenous people, in the sense that they had long settled in different parts of the Indian subcontinent before the Aryan invaded India from the North-west.

The concept of 'original inhabitant' is directly related to the initial peopling of India. Some anthropologists hypothesize that the region was settled by multiple human migrations over tens of millennia, which makes it even harder to select certain groups as being truly aboriginal. One narrative, largely based on genetic research, describes Negritos, similar to the *Andamanese* tribes of today, as the first humans to colonize India, approximately 30-65 thousand years ago. Sixty percent of all Indians share the mitochondrial DNA (mtDNA) haplogroup M, which is universal among Andamanese islander adivasis and might be a genetic legacy of the postulated first Indians. Some anthropologists theorize that these settlers were displaced by invading Austro–Asiatic-speaking Australoid people (who largely shared skin pigmentation and physiognomy with the Negritos, but had straight rather than curly hair), and STs such as the Irulas trace their origins to that displacement. The *Oraon adivasi* tribe of Eastern India and the *Korku* tribe of Western India are considered to be examples of groups of Australoid origin. Subsequent to the Australoids, most anthropologists and geneticists agree that Caucasoids (including both Dravidians and Indo–Aryans) and Mongoloids (Sino–Tibetans) immigrated into India: the Dravidians possibly from Iran, the Indo–Aryans possibly from the Central Asian steppes and the Tibeto–Burmans possibly from the Himalayan and Northeastern borders of the subcontinent. None of these hypotheses is free from debate and disagreement.

Ethnic origins and linguistic affiliations in India match only inexactly, however; while the *Oraon* are classified as an Australoid group, their language, called *Kurukh*, is Dravidian. *Khasis* and *Nicobarese* are considered to be Mongoloid groups and the *Munda* and *Santals* are Australoid groups, but all four speak Austro–Asiatic languages. The *Bhils* and *Gonds* are frequently classified as Australoid groups, yet Bhil languages are Indo–European and the *Gondi* language is Dravidian. Also, in post-colonial India, tribal languages have suffered huge setbacks with the formation of linguistic States after 1956 under the States Reorganization Act. For example, under State-sponsored educational pressure, *Irula* children are being taught *Tamil* and a sense of shame has begun to be associated with speaking the *Irula* language among some children and educated adults. Similarly, the *Santals* are 'gradually adopting languages' of the areas inhabited such as *Oriya* in Odisha, *Hindi* in Bihar and *Bengali* in West Bengal.

1.4.1. Pre-Mughal Period

Although considered uncivilized and primitive, tribals were usually not held to be intrinsically impure by surrounding (usually Dravidian or Aryan) caste based Hindu populace, unlike *Dalits,* who were. Thus, the tribal origins of *Valmiki,* who composed the Ramayana, were acknowledged, as were the origins of *adivasi* tribes such as the *Grasia* and *Bhilala*, which descended from mixed *Rajput* and *Bhil* marriages. Unlike the subjugation of the *Dalits*, the *adivasis* often enjoyed autonomy, and depending on region, evolved mixed hunter-gatherer and farming economies, controlling their lands as a joint *patrimony* of the tribe. In some areas, securing approval and support from the tribe was considered crucial by local rulers, and larger adivasi groups were able to sustain their own kingdoms in Central India. The *Gond*

Rajas of Garha–Mandla and Chanda are examples of an *adivasi* aristocracy that ruled in this region, and were 'not only the hereditary leaders of their Gond subjects, but also held sway over substantial communities of non-tribals who recognized them as their feudal lords'.

1.4.2. Mughal Period

This relative autonomy and collective ownership of tribal land by tribals was severely disrupted by the advent of the Mughals in the early sixteenth century. Rebellions against Mughal authority include the Bhil Rebellion of 1632 and the *Bhil–Gond* insurrection of 1643, which were both crushed by Mughal soldiers.

1.4.3. British Period

From the very early days of British rule, the tribesmen resented the British encroachments upon their tribal system and also their *raja* too did not welcome the British administrative innovations. They were found resisting or supporting their brethren of Tamar and Jhalda in rebellion. Beginning in the eighteenth century, the British added to the consolidation of feudalism in India, first under the *jagirdari* system and then under the *Zamindari* system. Beginning with the permanent settlement imposed by the British in Bengal and Bihar, which later became the template for a deepening of feudalism throughout India, the older social and economic system in the country began to alter radically. Land, both forest areas belonging to tribals and settled farmland belonging to non-adivasi peasants, was rapidly made the legal property of British – designated *Zamindars* (landlords), who in turn, moved to extract the maximum economic benefit possible from their new-found property and subjects. Lands of the tribals sometimes experienced an influx of non-local settlers, often brought from far away by the *Zamindars* to better exploit local land, forest and labour. Deprived of the forests and resources they traditionally depended on and sometimes coerced to pay taxes, many tribals were forced to borrow at usurious rates from moneylenders, often the *Zamindars* themselves. When they were unable to pay, that forced them to become bonded labourers for the *Zamindars*. Often, far from paying off the principal of their debt, they were unable even to offset the compounding interest, and this was made the justification for their children working for the *Zamindar* after the death of the initial borrower. In the case of the *Andamanese* tribe, long isolated from the outside world in autonomous societies, mere contact with outsiders was often sufficient to set off deadly epidemics in tribal populations, and it is alleged that some sections of the British government directly attempted to destroy some tribes.

Land dispossession and subjugation by British and *Zamindar* interests resulted in a number of *adivasi* revolts in the late eighteenth and early nineteenth centuries such as the *Santal hul* (or Santhal rebellion) of 1855-56. Although these were suppressed ruthlessly by the governing British authority (the East India Company prior to 1858, and the British government after 1858), partial restoration of privileges to *adivasi* elites (e.g., *Mankis*, the leaders of *Munda* tribes) and some leniency in tax burdens resulted in relative calm, despite continuing and widespread dispossession, from the late nineteenth century onwards. The economic deprivation, in some cases, triggered internal *adivasi* migrations within India that would continue for another century, including as labour for the emerging tea plantations in Assam.

1.4.4. Participation in Indian Independence Movement

There were tribal reform and rebellion movements during the period of the British Empire, some of which also participated in the Indian independence movement or attacked mission posts. There were several tribes in the Indian independence movement including Dharindhar Bhuyan, Laxman Naik, Jantya Bhil, Bangaru Devi and Rehma Vasave. During the period of British rule, India saw the rebellions of several backward castes; mainly tribals that revolted against British rule. These are: Great Kuki Invasion of 1860s, Halba rebellion (1774-79), Chakma rebellion (1776-1787), Chuar rebellion in Bengal (1795-1800), Bhopalpatnam Struggle (1795), Khurda Rebellion in Odisha (1817), Bhil rebellion (1822-1857), Paralkot rebellion (1825), Tarapur rebellion (1842-54), Maria rebellion (1842-63), First Freedom Struggle (1856-57), Bhil rebellion, begun by Tatya Tope in Banswara (1858), Koi revolt (1859), Gond rebellion, begun by Ramji Gond in Adilabad (1860), Muria rebellion (1876), Rani rebellion (1878–82), Bhumkal (1910), The Kuki Uprising (1917-1919) in Manipur, first Rampa Rebellion (1879), Vizagapatnam (now Vishakhapatnam District), second Rampa Rebellion (1921-1923), Vishakhapatnam District, Santhal Revolt (1885-1886), Munda rebellion.

1.5. MEANING OF RACE

People occupying different parts of the world differ from each other. There is variation in colour, facial features and stature. These differences have created the concept of race. Before the advent of modern scientific enquiry, the concept of race was used to differentiate people from each other. On this basis, the world population has been grouped into number of racial stocks according to environmental conditions and racial traits. No race can be said to be pure. A lot of intermixing has taken place between various races in the course of time. Race is a group of intermarrying people who are born of a common ancestor; possess similar physical traits, and similar we-feeling.

- According to **E.A. Hoebel**, race is biologically-inherited group, possessing a distinctive combination of physical traits that tends to breed true from generation to generation.
- As per **UNESCO**, race is a group, which shows in common a certain set of innate physical characters and geographically origin within a certain areas.
- According to **A.l. Krober**, a race is a valid biological concept. It is a group united by heredity, a breed or genetic strain or subspecies.
- As per **Frang Boas,** in common parlance when we speak of race, we mean a group of people that have certain bodily and perhaps also mental characteristics in common.
- **Gillen and Gillen** have given the meaning, as we may think of races, as group of mankind sharing a certain range of genotypes and distinguished thereby from other groups.

According to **D.N. Majumdar**, it is a group of people who, by their possession, have number of common physical traits that can be distinguished from others. Even if the members of this biological trait are widely scattered, they form the race.

1.6. CHARACTERISTICS OR TRAIT OF RACE OR CONCEPT CRITERIA FOR RACIAL CLASSIFICATION

The bases for classification of different races are:

- Form, colour, hair distribution on head, face, or body
- Body construction, stature, size of chest, and shoulders
- Form of skull, its height and width
- Facial characteristics such as nasal form, lip form, eye lid, ear, jaw, etc.
- Complexion of skin and eyes
- Length of arms and legs
- Blood type

A combination of these physical traits determines a race physically. The trait may be hereditary or environmental modification.

1.7. MAJOR RACES OF INDIA OR TYPES OF RACES IN INDIA

The latest racial classification of the tribal community of India has been made by Dr. B.S. Guha, which is based on anthropometric measurements carried out by him. The major races of India are discussed below.

1.7.1. Dravidian or Negrito Strain

It is regarded as the earliest strain found in India and elements of this strain are found among the tribes of Southern India. Physical characters of these strains are medium or long head, nose broad and flat, dark colour, hairs brizzy e. g., *Jarwar*, *Andamanese*, and *Kadar*. These broad-headed people from Africa were the earliest to inhabit India. They can still be found in the Andamans, and among hill tribes in South India such as the *Irulas*, *Kodars*, *Paniyans* and *Kurumbas*.

1.7.2. Pro–Australoid or Australian Strain

These are found among the tribes of central India and this strain numerically commands the biggest group in India. Physical characters of these strains are short and medium height, broad and flat nose, everted lips, and colour – medium to dark e.g., *Munda*, *Santhal*, *Korku*, and *Gond*.

1.7.3. Mongoloid Strain

They are evident among the tribes of Northeastern India with typical features of the eye and face. Generally, they are people of yellowish complexion, oblique eyes, high cheekbones, sparse hair and medium height. This strain is divided into two sub types - Paleo–Mongoloids and Tibeto–Mongoloids.

Paleo–Mongoloids

This racial type is further divided into two types:

1. ***Long headed*** - The *Angami Nagas* are of the long-headed type of the Paleo–Mongoloid race.
2. ***Broad headed*** - The people inhabiting in the Himalayan foot hills from Kashmir to Assam are said to be the broad-headed type of the Paleo–Mongoloid race.

Tibeto–Mongoloids

Physical characters of these strains are light skin colour, straight and scanty hairs on body, medium height, flat face and flat nose. The people of Sikkim and Bhutan are said to belong to the Tibeto–Mongoloid branch of Mongoloid race. e.g., *Naga*, *Lepchas*, and *Bhutias*.

1.7.4. Other Strains

1. The Mediterranean: (a) Paleo–Mediterranean, (b) Mediterranean, (c) Oriental
2. Western Brachycephals: (a) Alpenoid, (b) Dinaric, (c) Armenoid
3. Nordic Race

The first three races such as the Negrito, the Proto–Austroloid and the Mongoloid constitute the tribal population. The other three, namely the Mediterranean, the Western Brachycephals and the Nordic races constitute the general population of India.

1.8. MAJOR RACES OF THE WORLD

1.8.1. Negroid Race

People living in Africa, Arabia, Central and South India, Malaya, Australia and Tasmania belonged to the Negroid Race.

1.8.2. Pigmies

Seen in Africa, South-east Asia and Indonesia; it is a subgroup of the Negroids.

1.8.3. Mongoloid Race

Majority of the modern human race belongs to the Mongoloid race i.e., Japan, China, Indonesia, Formosa, Tibet, etc.

1.8.4. Hemitic Race

People living in ancient Egypt belonged to the Hemitic race.

1.8.5. Semitic Race

Ancient Babylonians, Hebrews, Phoenicians, Arabs, etc. belonged to the Semitic Race.

Note: The twentieth century racial classification by American anthropologist Carleton S. Coon divided humanity into five races; Caucasoid race, Congoid race, Capoid race, Mongoloid race, Australoid race. Semitic, Hemitic, and Indo–Europeans are the subgroups of the Caucasoid race.

Fig. 1. *Banjara* Tribe - Dance Culture

Fig. 2. *Kokna* Tribe - Culture, lifestyle and the unique mask for dance

2

TRIBAL: FAMILY AND MARRIAGES

2.1. MEANING OF TRIBAL FAMILY

Tribal family is a unique and basic unit of social institution/organization, which has no substitute. It is the nucleus of all social structure and still continues to be the most stable association and institution of human society. It plays a very important role in the development of personality of an individual and also in the process of socialization. A family is a small biological unit, which consists of father, mother and their children who are related with each other by kinship ties on the basis of marriage, blood or adoption.

The term 'family' has been derived from the Roman word 'Famulus', which means a servant. Family is defined as a group of people united by bonds of marriage, blood or adoption, interacting with each other in their respective social role of husband and wife, mother and father, son and daughter, brother and sister, and creating and maintaining a common culture. Family probably has grown out of the need to satisfy biological functions such as mating and protection of expectant mothers and infant child.

2.2. FUNCTIONS OF TRIBAL FAMILY

2.2.1. Biological Functions

This is an essential function, which the family performs. The sex instinct is the natural and biological urge of human beings. The satisfaction of sex desire requires that male and female should live together as husband and wife. It is most important and includes satisfaction of sexual desires of tribal husband and wife, reproduction, etc.

2.2.2. Psychological Functions

These functions include the protection and rearing of children, safety and security of wife, etc. The family is not only the birth place, but also the place of protection and care of children. The family provides daily care and personal protection to its dependant members namely aged, children, etc. The family is insurance for the individual in times of crisis, and it provides protection and shelter to orphans, widow and her children.

2.2.3. Economic Functions

Economic functions include fulfilment of economic needs such as food, clothing and shelter of its members. It provides the unit for economic pattern of the tribe and includes division of labour. On the basis of age, health and sex, the work has been divided among the family. The chief of family has to arrange for the livelihood of the whole family. Generally, children help mother and father for smaller work, women work in agriculture and cooking of food, and males are engaged in hunting and other economic activities to earn livelihood of family.

2.2.4. Social Functions

Social functions represent the establishment of a family status in society and its maintenance, socialization of children for the development of personality and their education. Through this process of socialization, the members learn their culture because the family moulds the behaviour of the members according to the culture. Thus, family is an important agency through which culture is transmitted to the present generation.

2.2.5. Religious Functions

A religious functions includes family *pooja*, or worship, visit to religious places, etc. The family provides religious training to its members. Children learn about their customs, traditions, taboos, art, music and dancing from the family.

2.2.6. Recreational Functions

The family provides recreational functions to its members such as visit to theatre, park, entertainment, dancing, singing, and other activities on ceremonial or special occasion. The members of the family also visit their relatives and enjoy various occasions in the family jointly and derive pleasure.

2.2.7. Health-related Functions

The family looks after the health of its members. It also gives them the knowledge about tribal system of medicine, which is mostly based on herbs. However, they have faith in magic, witchcraft, sorcery, etc. for cure of the disease. Tribals believe that diseases are caused by the bad spirits.

2.2.8. Organizational Functions

The tribal family happens to be the principal unit in the village organization. Each family contributes money, labour and personnel for the participation in the welfare of the village.

2.2.9. Educative Functions

Family is the first school of the children. The child learns the first letters under the guidance of parents. In the words of Mazzin, the first lesson of child begins between mother's kiss and father's care. The child learns language, behaviour and manners from the parents. The virtues of love, co-operation, obedience, sacrifice and discipline are learnt by the child in the family.

2.3. CHARACTERISTICS OF TRIBAL FAMILY

- Family has limited size and common language.
- Family members are emotionally attached.
- Husband and wife are permanent sexual partners.
- At least one member of the family has to be a bread earner for all the members.
- Family has a common nomenclature.
- Family is the cradle of education for all the children.
- Family inculcates sense of responsibility amongst its members.
- Family is the centre of education for the children.

2.4. TYPES OR FORMS OF TRIBAL FAMILY

Though family is an universal institution, its structure or forms vary from one society to another. Sociologists and anthropologists have mentioned about different types of families found in different cultures. Classification of families is, generally done on the basis of organization, forms of marriage, authority and residence. The classification/forms of tribal families are as follows:

2.4.1. Based on One Self

Family of Origin

A family in which one is born.

Family of Procreation

A family, which one sets up after marriage.

2.4.2. Based on the Way People Marry or Forms of Marriage

Monogamous

A monogamous family consists of one husband and one wife. In this type of family, one man has one wife or one woman has one husband at a given time. Hence, a husband and a wife living together constitute a monogamous family. It is an ideal form of family widely prevalent and this general pattern is found in all societies.

Polygamous

In polygamous family, a man has more than one wife. Generally, this type is found among the primitives, and a man, who has many wives, has great prestige. In such families, women do not object to their husband's taking on additional wives. This practice was popular among *Bhil* tribals in Central India. The relation between co-wives is usually friendly, but competition for the husband's attention is possible.

Polyandrous

In this type of family, a woman has more than one or several husbands. In South India, the *Todas* of the Nilgiris, *Khasa* of Jaunsar and *Bawar* of Uttar Pradesh practice polyandry. This is due to shortage of women as a consequence of female infanticide. Generally, in a family with three or four brothers, the brothers live together and share their wife without friction or jealousy.

2.4.3. Based on Extent of Family or Organization

Conjugal

Husband and wife.

Primary

Husband, wife and children.

Joint Family

Conjugal or primary family together with other relatives or any members.

Consanguineous

Conjugal or primary family together with blood relatives of husband or wife.

2.4.4. Based on Nature of Residence to which the Couple Moves After Marriage

Patrilocal Residence

This is a kind of family wherein a wife/married couple after marriage lives in her husband's family residence. Most of the tribal families are patrilocal and the descent is also traced through the male line. Mention may be made in this connection of the *Kharia*, *Ho*, *Bhil* and *Gond*. This is the general pattern of family found all over the country.

Matrilocal Residence

It is a family in which the married couple resides with the wife's family or kin group after marriage. In such families, the husbands either visit their wives periodically or live permanently with the matrilocal family. The women do not leave their families after marriage to live with their husbands. In communities where the system of matrilocal family prevails, the daughter is greatly valued. She will always remain with the family and will not be lost at marriage, whereas the boys will go to live with the families of their wives.

In India, there are some tribal groups, which have matrilocal families. The *Khasi* tribes of Assam found in the Jayantia hills have a tradition of matrilocal family. Similar primitive groups are also found in the Southern parts of the country. At the time of marriage, the *Khasia* are matrilocal, but afterwards when the husband becomes self sufficient he may set up his own residence.

Neolocal Residence

When the couple moves to an independent residence after marriage, which is neither patrilocal nor matrilocal residence, it is known as neolocal residence.

Avunculocal Residence

In this type of family, the married couple moves to the house of the maternal uncle and lives with his son after marriage. Avunculocal family is found among the *Nayars* of Kerala.

Matri-patrilocal Residence

In matri-patrilocal family, immediately after marriage, the bridegroom moves to the house of the bride and temporarily settles there till the birth of the first child. He then comes back to his family of orientation, along with his wife and child for permanent settlement. The *Chenchuas* of Andhra Pradesh live in this type of family.

2.4.5. Based on Nature of Ancestry and Property Inheritance

Patrilineal Family

When the descent is traced through the father, it is called patrilineal family. In this family, inheritance of property takes place along the male line of descent. The ancestry of such family is determined on the basis of male line or the father. After the death of the father, the eldest son is looked upon with respect since he represents his deceased father.

Matrilineal Family

In this type of family, descent is traced along the female line and inheritance of property also takes place along the female line of descent. The *Veddas*, the North–American Indians, some people of Malabar and the *Khasi* tribe are matrilineal. Generally, the matrilineal families are matrilocal.

2.4.6. Based on Inheritance of Name

Patronymic Family

When the children inherit the name of their father e. g., *Kharia* tribe.

Matronymic Family

When the children inherit the name of their mother e. g., *Khasi* tribe.

2.4.7. Based on Source of Flow of Authority

Patripotestal

Patripotestal family is a type of family in which all source of authority belongs to the paternal side. The oldest male or the father is the head of the family and he exercises his authority over the members of the family. The patrilineal family is also patrilocal and patripotestal. This is the general pattern of family found all over the world. In India, patripotestal families are found in *Gond*, *Santhal*, *Bhil* and *Ho* tribes. The tribals of the North-east also have patrilineal type of family. In this family, major decisions are taken by the male head and successor is also appointed by the male head of family.

Matripotestal

It is a form of family in which authority is centred in the wife or mother. The matripotestal family system implies rule of the family by the mother, not by the father. Women are entitled to perform religious rites and husband lives in the house of wife. It is also called mother-right family or maternal family under which the status, name and sometimes inheritance is transmitted through the female line. Such family is now found among the *Khasi* and *Garo* tribes of Assam and Meghalaya, among *Nayars* of Malabar in Kerala.

Avuncupotestal

When the source of authority is neither the father nor the mother, but may be uncle or other relatives.

2.5. TRIBAL MARRIAGE

India is a nation with vast and ancient traditions. The country blossoms with an extensive number of colourful cultures, languages, communities and tribes. Marriage, as a social institution, is an arrangement that enables persons to live together and co-operate with one another in an orderly social life and institutionalized relationship. It is essentially a re-arrangement of social structure, and India gives great significance to the tradition of marriage.

2.6. OBJECTIVES OF MARRIAGE

- Satisfaction of biological and physical need of husband and wife.
- Procreation.
- Satisfaction of social and economic needs.
- Satisfaction of religious needs.
- Providing healthy environment for growth and development of children and other family members.

2.7. CLASSIFICATION OR TYPES OF TRIBAL MARRIAGES

2.7.1. Monogamy

It is a form of marriage in which one man can marry with one woman at a time. It is the most common and acceptable form of marriage. A majority of the tribes in India practice monogamy. *Kamar* tribe of Madhya Pradesh commonly followed monogamy.

2.7.2. Polygamy

Polygamy means a system of marriage whereby a person marries to more than one person of opposite sex. Marriage to more than one person is fairly widespread all over the world. Among Indian tribes, there are many who are following this type of matrimonial alliance. Polygamy is classified into two groups as: polygyny and polyandry.

Polygyny

Polygyny means marriage of one man to several women. It is found among the *Naga*, *Gond*, *Baiga*, *Toda*, *Lushai* tribes and most of the Proto–Australoid tribes of Central India.

Polyandry

Polyandry refers to the marriage of one woman to several men. This type of matrimonial alliance is found in the Himalayan tract stretching from Kashmir to Assam and is popular among the Indo–Aryan and Mongoloid tribes. In its classical form, it is prevalent among the *Tiyan*, *Nishi* of Arunachal Pradesh, the *Theevas* and *Toda* of South India, the *Kota* and the *Ladakhi Bota*. It was also practiced by the *Khasa* tribe of Jaunsar region of Uttarakhand.

2.7.3. Hypergamy

Hypergamy is a marriage system in which the woman can marry with a man of a higher social class. In this marriage, a woman preferably marries a superior or marries an equal; however, a man should not marry a woman of higher status than himself.

2.7.4. Hypogamy

Hypogamy is the opposite of hypergamy. A man can select life partner above his caste and women can take husband from caste below her.

2.7.5. Exogamy

It is a rule of marriage in which the life partners are to be selected outside the group. Group may be caste, class, tribe, totem, family, race, village, religious group, etc. It is classified into three major types:

1. ***Family Exogamy*** - Marriage outside the family
2. ***Clan Exogamy*** - Marriage outside the clan
3. ***Tribe Exogamy*** - Marriage outside the tribe

2.7.6. Endogamy

It is a form of marriage in which the life partners are to be selected within the group. Group may be caste, class, tribe, totem, family, race, village, religious group, etc. It is classified into three major types:

1. ***Family Endogamy*** - Marriage inside the family
2. ***Clan Endogamy*** - Marriage inside the clan
3. ***Tribe Endogamy*** - Marriage inside the tribe

2.7.7. Preferential Marriage

In a family, sometimes certain sexual relations are permitted, which may not be inside the bond of marriage.

2.8. METHODS OF ACQUIRING MATES IN TRIBAL COMMUNITY

The following are the important methods of getting a mate in tribal society:

2.8.1. Probationary Marriage

The men and women are permitted to live together at the woman's house for quite some time. During this probation period, if they decide to marry then they can marry. In case the couple does not find each other's temperament to be suitable and compatible, they separate or if they do not decide to marry then the young man has to pay compensation to the girl's parents in the form of cash, kind or service. This system of marriage is found in *Kuki* tribe of Assam.

2.8.2. Marriage by Capture

An important method of securing a bride in Indian tribes is marriage by capture. This form of marriage is prevalent due to different types of causes. In the *Ho* tribe, its cause is the excessive bride price. Another cause is the misbehaviour to the woman by her sister-in-law; hence, that woman is always waiting for someone to carry her by force. The man captures the woman either by force or pre-arranged manner after which they can marry. The custom of marriage by capture is found among the *Santhal*, *Munda*, *Bhumij*, *Gond*, *Bhil*, *Kharia* and *Birhor* tribe of Bihar and *Naga* tribes. Among the *Bhumij* tribe, it is considered to be the best form of marriage.

2.8.3. Marriage by Trial or Test

Marriage by trial or test is the recognition of personal courage and bravery as highly desirable traits in a young man; and some tribes require a young man to prove his powers before he can claim the hand of any girl as his wife.

Fig. 3. Tribals on the way to market

Fig. 4. Tribal Family - Central India

Fig. 5. *Kolam* Tribe - Marriage ceremony

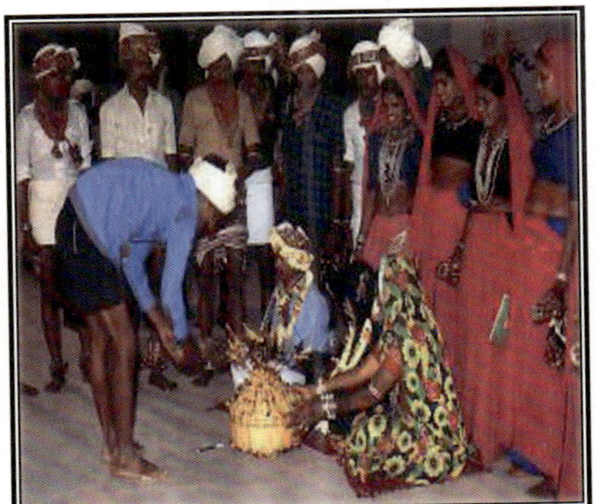

Fig. 6. Central India Tribes - Marriage procession

Fig. 7. *Baiga* Tribe - Brides

Fig. 8. *Baiga* Tribe - Grooms

Fig. 9. Tribal marriage in Jharkhand

Fig. 10. *Gond* Tribe - Brides

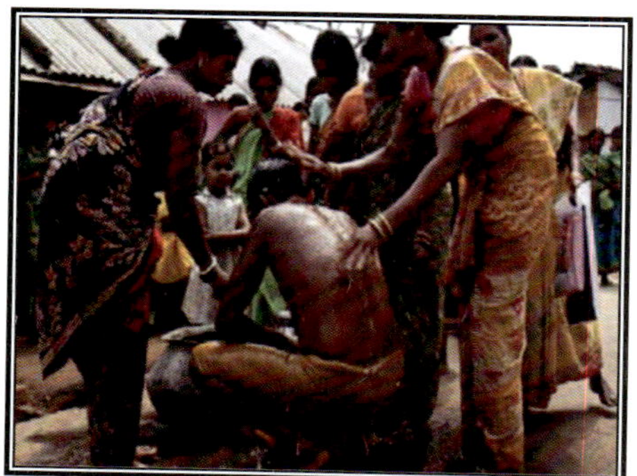

Fig. 11. Marriage Ceremony - Bathing to groom

Fig. 12. Northeastern tribe - Newly-married couple

In this form of marriage, a man is required to prove his courage and bravery by accomplishing the task assigned to him. The young man is allowed to marry the girl of his choice after he successfully passes the test of bravery. A strange custom of this type of marriage is practiced among *Bhil* tribe, which is known as *Gol Gadbedo*.

In this tribe, a social function is celebrated around a tree or a high pole at the top of which a coconut and some sweets are put. The boys and girls dance around this pole or tree, the girls remaining nearer the tree and the boys forming their circle outside the circle of the girls. Now a young man suddenly tries to break the circle of the girls and reaches the pole or tree, eats the sweets and breaks the coconut. The girls beat him and try to restrain him. If he comes out successfully, he is given the right to select any girl from the gathering as his wife.

2.8.4. Marriage by Purchase

The most prevalent method of securing bride in primitive societies is marriage by purchase. The bride's price is not paid as the price for the girl, but symbolizes her utility for the male and the compensation to the parents for giving away their daughter in marriage. The man pays a price for the bride to her parents before he has her as his wife. The price paid may be in the form of cash or kind or both.

2.8.5. Marrige by Service

Some men may not be economically sound to pay the bride's price. Then he has to pay the price by serving his would-be father-in-law to the equivalent of the bride's price, after which he can marry and take away the bride. In service marriage, the youth has to serve his future father-in-law for 3-5 years. During this period, if the father of the girl is satisfied with the youth, he gives his daughter in marriage to him. If, however, the father is dissatisfied, he dismisses the youth and gives this chance to another person. During the course of service, the young man can meet the girl, but cannot enter into sex relationship with her. This custom is particularly found among those persons who are too poor to pay any money as bride's price.

In *Gonds* and *Baigas* when the bridegroom is not in a position to pay the bride's price, he goes to serve in his would-be father-in-law's house as a suitor–servant, and marries the desired girl after which they both return to his household. In the *Gond* tribe, the bridegroom, desirous of marrying through service, is known as *Lamnai* and *Lamsena* in *Vaiga* tribe. In Himachal Pradesh, he is known as *Kamaoos*.

2.8.6. Marrige by Exchange

If a son and daughter of marriageable age are available in two families, they may exchange the daughters and thus, obtain bride for their son. In this type of marriage, when a man's son is prepared to marry another's daughter, the latter's son accepts the former's daughter in marriage; thus an exchange of boys and girls is made between the two families.

2.8.7. Marriage by Mutual Consent

An interesting form of securing a bride prevalent in Indian tribes is marriage by mutual consent and elopement. This was known as *Gandharva* marriage in the ancient Hindu scriptures. Among the *Ho* tribe of Bihar, it is known as *Rajikhushi* marriage. Mutual consent of the parents and their approval is required before marriage.

2.8.8. Marriage by Intrusion

This is a reverse case of marriage by capture. In this type of marriage, the girl forcibly enters the house without the permission of her lover or desires to marry with unwilling boy and does not leave inspite of bearing all types of insults and cruelties. She starts household work there such as sweeping the floor and washing the utensils. If she can withstand the abusing and the maltreatment of the mother and the sister of the boy for a specific period of time, then she will be taken as wife of that particular boy, and usual marriage function will be held later on. In this process, she is subjected to humiliating treatments, but afterwards is accepted. This custom of marriage is found among the *Santal*, *Oraon*, *Ho* and *Brishor* tribes.

3

TRIBAL : KINSHIP, TABOO, TOTEM AND YOUTH DORMITORY

3.1. MEANING OF KINSHIP

Kinship represents one of the basic social institutions. It is universal and in most societies plays a significant role in the socialization of individuals and the maintenance of group solidarity, and also provides continuity between generations. It is very important in primitive societies and extends its influence on almost all their activities. Kinship is a method of acknowledging relationship. It is a social bond initiated by genetic or blood ties. Marriage kinship ties are of the fundamental importance in every society all over the world.

Man is social by nature. He establishes many types of relationships with a number of persons. You find yourself surrounded by several kins in your family as well as outside it. In all, society people are bound together in groups by various kinds of bonds. Kinship is an interaction with the members of society and its relationship through marriage, blood, friendship, neighbour, enemies, etc. This bond is known as kinship. The most universal and most basic of the bonds or relationships is that, which is based on reproduction i.e., bond of marriage or blood. The kinship among the tribal society can be traced with the help of Fig. 13.

Kinsmen – Male relatives, Kinswomen – Female relatives.

3.2. TYPES OF KINSHIP

3.2.1. Affinal Kinship

Affinal kinship is based on marriage. The most primary affinal relationship is the one between a husband and a wife, which in its extended form includes parents and siblings of both sides, and their spouses and children. The relationship or bond arising out of socially or legally-accepted marriage is called affinal kinship. New relations are created when marriage takes place. Not only man establishes relationship with the girl and the members of her family, but also family members of both the man and the woman get bound among themselves. This is the bond or relationship between the husband and wife, and their close relatives. Some of the examples of affinal kinship are: brother-in-law, sister-in-law, father-in-law, mother-in-law, son-in-law, daughter-in-law, etc.

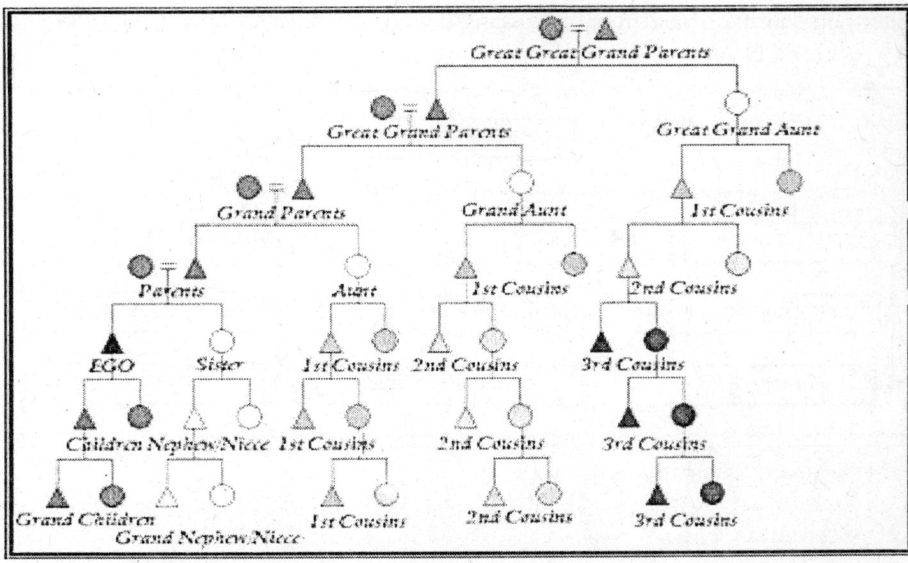

Fig. 13. Flow of kinship in tribal society

3.2.2. Consanguineous Kinship

Consanguineous kinship is based on descent; commonly known as blood relation. The relationship or bond arising due to blood connections is called consanguineous kinship. This is the bond between parents and their children, and among the children of the same parents. The examples of consanguineous kinship are: father–son, father–daughter, mother–daughter, mother–son, brother–brother, brother–sister, sister–sister, etc. Blood relationship or consanguineous kinship is based on biological fact of genetic relationship between parents and children. The relationship between mother and child is the starting point of consanguineous kinship, which in its extended form includes the child's father, grandparents, uncles, cousins, aunts and so on.

3.3. DEGREE OF KINSHIP

The kinship relationship can also be categorised in terms of degree of closeness into following types:

3.3.1. Primary Kins

Primary kins or first degree kins are those who are directly related to each other. These kins may belong to the family of orientation i.e., the one in which we are born or brought up (Fig. 14). For example, our father, mother, brothers, and sisters are our primary kins because we are directly related to these persons through blood or genetic ties. Husband–wife relationship is an example of primary affinal kin. A person may have seven types of primary kin viz., mother, father, son, daughter, brother, sister, husband or wife. In primary kinship a person is directly related to another person either by affinal or consanguineous bond. If X is the primary kin of Y then Y is also a primary kin of X.

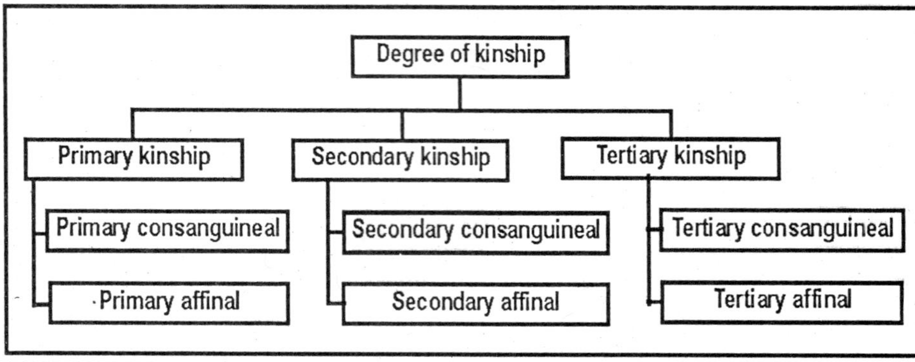

Fig. 14: Degree of kinship in tribal society

3.3.2. Secondary Kin

The primary kins of our first degree kins are our secondary kins. They are not directly related to us, but through primary kins. There can be 33 types of secondary kins. Some of the relations of secondary kins are: Father's father i.e., paternal grandfather, mother's father i.e., maternal grandfather, father's mother i.e., paternal grandmother, wife's or husband's brother and sister, parents and so on (Fig. 14).

3.3.3. Tertiary Kin

The primary kins of our secondary kins or secondary kins of our primary kins are known as our tertiary kins or third degree kins (Fig. 14). For example, your brother-in-law is your secondary kin, and his wife or children who are his primary kins become your tertiary kin. There are 151 types of tertiary kins. For example, father's brother's wife, father's sister's husband, father's mother's brother, father's mother's sister and so on.

3.4. KINSHIP BASED ON DESCENT

3.4.1. Patrilineal Kin

Patrilineal refers to tracing one's ancestry through the paternal line, meaning the father's line. Relatives are identified by tracing descent exclusively through males from a founding male ancestor. In patrilineal kins, the descent of power and property takes place through the male line and girls are raised for other families. It refers to a group of people whose common ancestor is a male.

3.4.2. Matrilineal Kin

In matrilineal kins, the relatives are identified by tracing descent exclusively through females from a founding female or mother and maternal ancestors. In this, the children belong to the mother's descent group, the inclusion of a husband in the household is less important and women usually have higher status. It refers to a group of people whose common ancestor is a female.

3.5. KINSHIP USAGES OR BEHAVIOUR

Kinship usages or the rules of kinship are significant in understanding kinship system. They serve two main purposes: They create groups or special groupings or kin, and kinship usage provides guidelines for interaction among persons in these social groupings. It defines proper and acceptable relationships. Thus, it acts as a regulator of social life. Kinship behaviour or usages refers to definite and comparatively-stable patterns of behaviour of different members of a kin group. These behaviour patterns may be verbal and/or non-verbal. Some of the kinship usages are discussed below:

3.5.1. Avoidance

It means that two kins, normally of opposite sex, should avoid each other. In almost all societies, avoidance rules prescribe that men and women must maintain certain amount of modesty in speech, dress and gesture in a mixed company. Thus, a father-in-law should avoid daughter-in-law. The *Purdah* system in family in the North illustrates the usage of avoidance. Certain relatives avoid talking to each other directly, avoid physical contact and maintain minimum social interaction with each other and so on. Some of the kins covered by such restrictions are – parents-in-law and daughter-in-law, mother-in-law and daughter-in-law, mother-in-law and son-in-law, husband's elder brother and younger brother's wife, etc. Some of the examples of avoidance are:

- The Hindu wife is not supposed to call her husband by his name.
- Amongst the *Yukafir*, the son-in-law is not supposed to see the face of his mother-in-law and father-in-law.
- Amongst the *Ostiyak*, the married man is not supposed to see the face of his mother-in-law at least till he gets a child.
- Amongst the *Aruntas*, if the mother-in-law enters or approaches the hut of her son-in-law, she would be excommunicated.
- Amongst the *Veddas* of Ceylone, elder brother and sisters are not supposed to live in the same house and even eat together.
- Among the tribes of Andaman and Nicobar Islands, the elder brother does not talk with his younger brother's wife.

The avoidance may be due the following reasons:

Existence of Matrimonial System

In the past, the matrimonial systems were in existence. The husband had to stay at his wife's residence. He used to be a stranger in that family, and used to be kept away from his parents. After marriage, women do not know the social and cultural value of husband's house.

To avoid Incest Taboos

In certain tribes, even brother and sister were not allowed to stay under one roof to avoid incest taboo.

To avoid Conflict

Whenever people come in contact with each other, the co-operation and conflicts are always equally present. But in certain tribes, hostility is regarded against social norms hence, putting restrictions help to avoid conflicts.

3.5.2. Teknonymy

According to the usage of this, a kin is not referred to directly, but is referred through another kin. In a traditional Hindu family, wife does not directly utter the name of her husband, but refers to her husband as 'the father of so and so'.

3.5.3. Avunculate

It refers to the special relationship that persists in some societies between a man and his mother's brother. This usage is found in a matriarchal system in which prominence is given to the maternal uncle in the life of his nephews and nieces. The maternal uncle is considered more important than even his father. This is common in matrilineal family.

3.5.4. Amitate

The usage of amitate gives special role to the father's sister. Here, father's sister is given more respect than the mother. Among *Todas*, the child gets the name not through its parents, but through the father's sister. Naming the child is her privilege. This usage is normally prevalent in patrilineal systems.

3.5.5. Couvade

This kinship usage involves only husband and wife. According to this usage, the husband is made to lead the life of an invalid along with his wife whenever she gives birth to a child. He is then not supposed to engage himself in hard work, but is expected to observe dieting and certain other taboos. According to Malinowski, the usage of couvade contributes to a strong marital bond between the husband and wife. The usage of couvade is prevalent among the *Khasi* and the *Todas* tribes.

3.5.6. Joking Relationship

It is just opposite of avoidance. A joking relationship involves a particular combination of friendliness and antagonism between individuals and groups in certain social situations. The relatives are free to crack jokes at each other's expense, tease each other and make fun of each other under this form of behaviour. The relatives under this category are expected not to take offence or to mind each other's conduct i.e., joking relationship of *Bhabhi* and *Dewar*, wife's sister and brother, husband and wife, etc. Some of the examples are:

- Amongst the *Oraons* of Odisha and the *Baigas* of Madhya Pradesh, such relationships prevail between the grandfather and grandmother, and their grandchildren. Majumdar and Madan have cited the example of a case in which a grandfather had married his granddaughter and got a child from her.
- Amongst the Crow–Indians, such relationships may prevail between a man and his wife's sisters. They could be very friendly and even talk freely about sex matters.

3.6. MEANING OF TABOO

Each society is characterized by its own belief system that takes roots from its cultural identity. This belief system affects all factors of human life including health, ritual, marriage, caste, culture, social, religion, conversation, etc. Tribals are no exception to it. The tribal societies of India have their distinctive sets of taboo and socio-cultural practices. Totemism and taboo; both are intimately connected with the history of early religion. The two lines of development, totemism and taboo, appear side by side in early cults, and influence each other; but their functions in the social organization of religion have been different, and they are best treated separately.

Taboo means to forbid or forbidden. D. N. Majumdar says that taboo is used to designate all the restrictions communicated through verbal 'do' or 'do not', and is generally associated with ritualistic behaviour. Basically, it is an unwritten law of tribal society and whose violation is not dealt with by the society. The violator of taboo becomes its culprit and is punished by the tribal chief.

Many taboos popular in the primitive society continue to be observed. These are prohibitions and restrictions that were forced upon the primitives. The breach of a taboo invites due consequences. The important taboos are related to puberty, menstruation and child birth. There were interesting social taboos. It was objectionable for a man to talk to his nieces after they had attained puberty. A wife was forbidden to address her husband by name. Among *Mannans, Muthuvans, Paniyans, Uralis* and some others, women were bounded to keep away from the presence of their father-in-law. *Uralikurumans* and *Visavans* had *Chavadies* (Common halls) for young men, which combined the features of a dormitory, club and guest room, so that they might have no contact with females and no knowledge of the sexual life within homes. There were taboos regulating diet, drinking, smoking and the use of vessels.

3.7. PURPOSE OF TABOOS

According to Majumdar and Madan, 'The woof and warp of magical beliefs and practices is made up of positive and negative strands, and taboo is later. Its purpose is three-fold, productive, protective and prohibitive'. The productive purpose of taboo is the growth of cultivation and other occupation. Its protective purpose is the safety of the individual, children, women, family and tribe. Sometimes, some person keeps them away from certain objects, actions or places. Its prohibitive purpose is to prohibit the members of tribe from doing specific activities. These include limited contact with magician or menstruating women.

Some of the examples of taboos are:

- Pregnant women should not cut the watermelon, pumpkin, or pluck the fruits of lemon.
- Pass under the tree of Peepal and Tamarind, and cut the trees from degraded land; violation can attract bad spirit.
- Tribals considered that cutting of Banyan and Peepal trees is more sinful than killing a saint.
- Cutting of timber after sunset is believed to attract illness of children.
- It is believed that the burning of Banyan, Peepal and Bamboo for burning in *Yagya* (holy fire) can destroy the whole family.
- Urinating under the Mango trees can cause leprosy.
- Among the *Bodo* of Assam, there is a taboo against marrying within the same clan.

3.8. MEANING OF TOTEM

An important feature of tribal religion is the custom of totem. The people belonging to different tribes believe in different types of totems. The totem is an object or an animal, which has a mystic relationship with the members of a tribe. The members of a tribe have much regard towards their totem. They cannot kill their totem if it is thought to be an animal. In this way, taboos are linked with the totems. In the opinion of E. A. Hoebel 'Totem is an object, often an animal or a plant, held in special regard by the members of a special group, who feel there is a particular bond of emotional identity between themselves and totem'.

Totemism is the fundamental basis of the religious organization among the tribals. It strengthens community consciousness and brotherhood among the members of the group. It acts as a symbol of group unity. It can undoubtedly be said that totem occupies an important place in tribal religion. A tribe finds its origin in an animal. Some tribes find their origin from some plant as well. Mystical animal from which a tribe finds its origin is its totem. This Totem is very sacred for the tribe. Its figure is usually inscribed on parts of body of the people belonging to that tribe. The tribe is not allowed to kill that animal except on very special and sacred occasions. It is believed that totem animal protects the tribes in all difficult situations and at all hard times. It helps in keeping the tribes in bonds of unity and bestows dignity on the tribe as a whole. It helps in maintaining consciousness and brotherhood. Totem animal is considered to be the beginner of life of the tribe. Totem used in tribal society is depicted in Fig. 15.

The *Birhor* tribe, in the Chota Nagpur district presents the illustration of strongly-formed patrilineal totemism. Among the *Oraon* of Chota Nagpur, 67 totems as derived from animals or plants have been recorded. The *Munda* speaking tribal group developed strong totemism from their earlier proto-totemic organization. Among the *Mundas*, the majority of the exogamous clans have animal or plant totem. There are in vogue other kinds of totems such as rainbow, umbrella or basket as well. Some of the totems used in tribes is given in Table 1.

Fig. 15. Totems used in tribal society

Table 1: Totems used in tribal community

Sr. No.	Surname	Tribe	Totem
1.	Baa, Khess, Xess	*Kharia, Oraon*	Paddy
2.	Bara	*Oraon*	Tree
3.	Barwa	*Oraon*	Wild boar
4.	Baxla	*Oraon*	Grass
5.	Beck, Bilung	*Oraon, Munda, Kharia*	Salt
6.	Dundung, Minj / Minz	*Kharia, Oraon*	Fish
7.	Ekka, Kachhap, Kullu	*Oraon, Kharia*	Tortoise
8.	Jojo	*Munda*	Tamarind

Sr. No.	Surname	Tribe	Totem
9.	Kerketta, Toppo, Tete	*Oraon, Munda, Kharia*	Bird
10.	Khakha, Xaxa	*Oraon*	Crow
11.	Kindo	*Oraon*	Fish
12.	Kiro, Lakra	*Kharia, Oraon*	Tiger
13.	Kispotta	*Oraon*	Intestine of a pig
14.	Kujur	*Oraon*	Creeper (Kujuri Phal)
15.	Soreng	*Kharia*	Stone
16.	Tigga	*Oraon*	Monkey
17.	Tirkey	*Oraon*	Tree-bandar Lowry
18.	Xalxo	*Oraon*	Sea fish

Source: Tribal Zone, 2014

3.9. MEANING OF YOUTH DORMITORY

The most important social institution in the tribal-social organization is youth dormitory, which is like a preparatory school for the tribal children. Dormitories are youth houses facilitating young people to come and live together, and are common among tribal communities. In India, dormitories have played an important role in the cultural, social and religious evolution of many tribes. Youth dormitories are confined to remote tribal belt of India including States of Kerala, Chhattisgarh, Odisha, Jharkhand, Maharashtra, Madhya Pradesh, Andhra Pradesh, Meghalaya, Assam, Tripura, Mizoram, Manipur, Nagaland, Arunachal Pradesh, and Andaman and Nicobar islands.

The English word 'dormitory' is derived from the Latin word dormitorium, which signifies a communal or collective sleeping quarter. These youth organizations are centred in big buildings known as dormitories. They are made up of straw, bamboo, grass and thatch. Youth dormitory is the largest hut in the village with three walls; profusely decorated with symbols representing animals, plants and tribal paintings. The fourth side is open. These organizations are found in almost every tribe and their social life is centred on it. In the evening, the members of the dormitory gather after taking their meals at their homes. At the dormitory, they gather after fire, sit around it, tell stories, sing, dance and play, and finally sleep. There are separate dormitories meant for boys and girls. All the youths of the tribe pass their night in the dormitory. In the villages where there is no dormitory for the girls, they sleep in the house of some old woman.

3.10. CHARACTERISTICS OF YOUTH DORMITORY

- While at some places there are separate dormitories for boys and girls, in other tribes they live in a common dormitory. Among the *Nagas* of Assam, the boys live in the *Morung* and the girls sleep at *Yo*. This is also common custom in *Muria* and *Ho* tribe.

- Normally, the dormitories are situated outside the village in the forest, but they may also be near the agriculture fields as found in the case of dormitories of the *Nagas*. In *Oraon* tribe, the dormitory is situated in the centre of the village.
- The life in the dormitory is based on certain traditions and customs, which are invariably followed by all the members.
- The age of the membership of the dormitory differs from tribe to tribe. As a general rule, it is four or five years in most of the tribes.
- The boys and girls live in the dormitory till their marriage. Their membership is automatically dissolved after marriage.
- If a girl becomes a widow, she can again enter the dormitory as its member without any prior permission.
- The members of the dormitory are divided into two classes according to age – seniors and juniors. The head of the dormitory is selected from the senior group. It is his job to take care of all the members of the dormitory and maintain discipline among them. The juniors follow the commands of the seniors and receive different types of education from them.
- It is the duty of every member of the dormitory to keep everything secret about the dormitory.
- The members of the dormitory carry out several functions together such as construction of a house on the occasion of marriage, work together in village common functions and helping the village folk in harvesting the crop.

3.11. FUNCTIONS/PURPOSE/IMPORTANCE OF TRIBAL DORMITORY

- To help in the gathering of edible things and consolidating economic organization.
- To educate boys and girls in social and other duties, and to impart them education in the matters of sex.
- To follow the principle of endogamy, and to keep the movement of women limited and controlled.
- According to Majumdar and Hutton, the dormitories are the centres of social education.
- Purpose of the dormitory is to keep boys and girls unaware of the sexual relationships of their parents.
- According to Majumdar, the purpose of dormitory is to gather the youths of the village at one place to save the village from the attack of wild animals and the village maidens from the attack of males of other tribes. Besides it, the origin of the dormitory may be due to want of houses, the bliss of communal life, and the development of tribal culture.

- In it are taught the lessons of the traditions, culture, norms, ideals, religious beliefs, methods of earning livelihood and discipline to the boys and girls of the tribe under specific rules and regulations.
- In it are shown the dances imitating agriculture, hunting, gathering honey and economic activities carried out by the village folk.
- The members of the dormitory carry out several functions together such as construction of a house on the occasion of marriage, co-operation in some common village functions, and helping the village folk in harvesting the crop.
- The conceptions of crime and punishment, and good and bad actions are taught through different stories.

3.12. THE *GHOTUL* OF *MURIA* AND *GOND*

Among the dormitories of Indian tribe, the *Ghotul* of the *Muria* and *Gonds* of Central India are worth mentioning. It is raised on the bamboo poles and the roofs of straw. There is no door or window in it. The members enter through a small hole sufficient to permit entry. All the unmarried boys and girls gather here in the evening, make a fire, sing and dance, tell traditional legends and sleep here. The juniors serve the seniors and receive different types of training from them.

Among the members of the *Ghotul*, different boys and girls are given different status and the corresponding roles. The head of the dormitory is known as *Challen* or *Silledar*. Under him, there are *Dewan*, *Tehsildar*, *Subedar* and *Kotwal*. Similar positions are also given to the girls. According to Majumdar, the purpose of the *Ghotul* is to save the village folk from the attack of the wild animals. The youths remain awake for most of the night and sleep only in its last lap or in the day. They also help their parents in the field or in grazing the cattle in the forest.

Among the *Muria* tribals, the girls massage the bodies of the boys. Verrier Elvin has pointed out that among the *Muria* tribals, the senior girls train the junior boys in matters of sex.

3.13. DORMITORIES FOR OTHER TRIBES

Dormitories are known by different names in different tribes – among the *Nagas* of Assam, they are called *Morung*, while the female youth dormitory is called *Yo*, and among *Angomi Nagas* it is called *Kinchuki*. In Uttar Pradesh, it is known as *Rangbang*, while among the *Mundas* and *Ho* tribes of Central India, it is called *Gitoria*. The *Oraons* call it *Dhumkuriya*, the *Bhuiyan* call it *Dhangarbassa*. In Odisha, this institution occurs among many tribal communities in same form or the other. The *Juangs* call it *Majang* and *Darbar*, the *Kondhs* call it *Dindaghar* and among the *Bondos* it goes by the name *Ingersin*.

The *Oraon* youth has to pass his night in the *Dhumkuriya* situated outside the village. In the *Ho* tribe, the unmarried boys and girls live in separate dormitories. Among the *Lhota Nagas* of Assam, the boys can leave the *Morung* only in the case of serious illness of their mothers. Among the *Miami* tribals of Assam, both boys and girls sleep in the same dormitory; the girls sleep on the ground floor, the boys sleep on the first floor. Similar is the custom in the *Muria* tribe of Bastar region of Chhattisgarh. Their dormitories are situated outside the village, and the boys and girls of the age 15 and 16 sleep during night. The unmarried boys and girls gather here at night, sing and dance, and participate in other social activities and finally pass the night. Such is the custom in the *Munda* tribe also. Distribution of youth dormitories is shown in Table 2.

Table 2: Distribution of youth dormitories in India

S.N.	State/Union Territory	Tribes	Locations
1.	Andaman and Nicobar	*Jarawa*	Andaman
2.	Arunachal Pradesh	*Adi, Wancho*	East, West Siang, Tirap
3.	Assam	*Boda, Kachari*	Darrang, Goalpara
4.		*Karbi*	Mikir and N.C.Hills
5.		*Deueri*	Lakhimpur, Sibsagar
6.		*Lalung*	Nowgaon Lakhimpur
7.	Chattisgarh	*Muria,*	Bastar
8.		*Abhuja Maria*	Bastar
9.		*Dorla*	Bastar
10.	Jharkhand	*Oraon*	Palamau, Ranchi
11.		*Munda*	Giridih, Ranchi
12.		*Birji*	Palamau, Ranchi
13.		*Birhor*	Hazaribagh, Ranchi
14.	Kerala	*Mujhuvan, Manna*	Trivendrum, Palghat
15.	Manipur	*Tangkhul, Paite*	South and East Manipur Hills
16.	Meghalaya	*Garo, Hajoung*	West and East Garo Hills
17.	Mizoram	*Lusei, Hmar*	Aizawl, Lunglei
18.	Nagaland	*Lohta, Rengma*	Wokha, Phek
19.	Odisha	*Juang, Santal*	Dhenkanal, Keunjhar, Baleswar, Mayurbhanj
20.	Tripura	*Durlong kuki*	North Tripura

Source: Roy Shibani, 2006.

Fig. 16. *Naga* tribe - Dance around *Morung* (Youth Dormitory)

Fig. 17. *Gond* Tribe - Dance around *Ghotul* (Youth Dormitory)

Fig. 18. One of the oldest youth dormitory in Nagaland

Fig. 19. Youth dormitory in West Bengal

Fig. 20. Youth dormitory of *Mizo* Tribe - *Zawlbuk*

Fig. 21. *Morung*-Youth dormitory of *Naga* Tribe

Fig. 22. *Ghotul* - Youth Dormitory of Central Zone tribes

Fig. 23. *Ghotul*-Tribal Hut of *Gond* and *Muria*

4

TRIBAL: DEMOGRAPHY AND DIALECTS

4.1. INTRODUCTION

India has the second largest concentration of indigenous population in the world. The tribal community in India have been living, generally in the remote hilly and plain forest areas. The tribal areas constitute a significant portion of backward areas around 18.74% area of the country. It is estimated that there are 300 million tribal people residing all over the world. Nearly half of them are living in China and India. In India alone, being a tropical country with rich biological and cultural diversity, there are more than 100 million tribals.

Tribal communities live in various ecological and geo-climatic conditions ranging from plains and forests to hills and inaccessible areas. Tribal groups are at different stages of social, economic and educational development. While some tribal communities have adopted a mainstream way of life, at the other end of the spectrum, there are certain STs, 75 in number, known as Particularly Vulnerable Tribal Groups (PVTGs), and are characterized by:

- Pre-agriculture level of technology
- Stagnant or declining population
- Extremely low literacy
- Subsistence level of economy

4.2. DISTRIBUTION OF TRIBAL POPULATION

The STs are notified under Article 342 of the Constitution of India, spread in 30 States/Union Territories and the number of individual ethnic groups notified as STs is 705. The tribal population of the country, as per 2011 Census, is 10.43 crore, constituting 8.6% of the total population. Of them, 89.97% live in rural areas and 10.03% in urban areas. The decadal population growth of the tribal's from Census 2001 to 2011 has been 23.66% against the 17.69% of the entire population. The trend in ST population since Census 1961 is illustrated in Table 3. The ST population has increased from 30.1 million in 1961 to 104.3 million in 2011. The sex ratio for the overall population is 940 females per 1,000 males and that of STs is 990 females per 1,000 males.

Table 3: Trends in proportion of ST population

Census Year	Total population (in millions)	STs population (in millions)	Proportion of STs population
1961	439.2	30.1	6.9
1971	547.9	38.0	6.9
1981 #	665.3	51.6	7.8
1991 @	838.6	67.8	8.1
2001 $	1,028.6	84.3	8.2
2011	1,210.8	104.3	8.6

Excludes Assam in 1981

@ Excludes Jammu and Kashmir in 1991

$ The figures exclude Mao–Maram, Paomata and Purul sub divisions of Senapati district of Manipur, Census 2001

Source: Census of India, 2011

Out of the total ST population of 104.3 million, maximum population of the tribe is found in Madhya Pradesh (14.7%) and minimum in Daman and Diu. Statewise lowest and highest tribal population is depicted in Table 6.

There is absence of tribes in National Capital Territory (NCT) of Delhi, Haryana, Chandigarh, Pondicherry/Puduchgerry and Punjab. In India, the details of the Statewise population of ST as per Census 2011 is given in Table 4, Figure 24 and percentage of tribals to the total population of State (Census 2011) is shown in Table 5. Distribution of *Gond* and *Pardhi* tribe in India is shown in Fig. 25 and Fig. 26.

Table 4: Statewise population of ST

Sr. No.	State/ Union Territory	Total ST population	Percentage to the total tribal population
1.	Madhya Pradesh	1,53,16,784	14.7
2.	Maharashtra	1,05,10,213	10.1
3.	Odisha	95,90,756	9.2
4.	Rajasthan	92,38,534	8.9
5.	Gujarat	89,17,174	8.6
6.	Jharkhand	86,45,042	8.3
7.	Chhattisgarh	78,22,902	7.5
8.	Andhra Pradesh	59,18,073	5.7
9.	West Bengal	52,96,953	5.1
10.	Karnataka	42,48,987	4.1
11.	Assam	38,84,371	3.7
12.	Meghalaya	25,55,861	2.5

Sr. No.	State/ Union Territory	Total ST population	Percentage to the total tribal population
13.	Nagaland	17,10,973	1.6
14.	Jammu and Kashmir	14,93,299	1.4
15.	Bihar	13,36,573	1.3
16.	Tripura	11,66,813	1.1
17.	Uttar Pradesh	11,34,273	1.1
18.	Mizoram	10,36,115	1.0
19.	Arunachal Pradesh	9,51,821	0.9
20.	Manipur	9,02,740	0.9
21.	Tamil Nadu	7,94,697	0.8
22.	Kerala	4,84,839	0.5
23.	Himachal Pradesh	3,92,126	0.4
24.	Uttarakhand	2,91,903	0.3
25.	Sikkim	2,06,360	0.2
26.	Dadra and Nagar Haveli	1,78,564	0.2
27.	Goa	1,49,275	0.1
28.	Lakshadweep	61,120	
29.	Daman and Diu	15,363	
30.	NCT of Delhi	Nil	-
31.	Haryana	Nil	-
32.	Chandigarh	Nil	-
33.	Puducherry	Nil	-
34.	Punjab	Nil	-
35.	TOTAL (INDIA)	10,42,81,034	

Source: Census of India, 2011

Table 5: Percentage of tribals to the total population of State (Census, 2011)

Sr. No.	State	Total population of State	ST population	Percentage of ST
1.	Lakshadweep	64,473	61,120	94.8
2.	Mizoram	10,97,206	10,36,115	94.4
3.	Nagaland	19,78,502	17,10,973	86.5
4.	Meghalaya	29,66,889	25,55,861	86.1
5.	Arunachal Pradesh	13,83,727	9,51,821	68.8
6.	Dadra and Nagar Haveli	3,43,709	1,78,564	52.0

Sr. No.	State	Total population of State	ST population	Percentage of ST
7.	Manipur (Excluding three Sub-divisions of Senapati District)	25,70,390	9,02,740	35.1
8.	Sikkim	6,10,577	2,06,360	33.8
9.	Tripura	36,73,917	11,66,813	31.8
10.	Chhattisgarh	2,55,45,198	78,22,902	30.6
11.	Jharkhand	3,29,88,134	86,45,042	26.2
12.	Odisha	4,19,74,218	95,90,756	22.8
13.	Madhya Pradesh	7,26,26,809	1,53,16,784	21.1
14.	Gujarat	6,04,39,692	89,17,174	14.8
15.	Rajasthan	6,85,48,437	92,38,534	13.5
16.	Assam	3,12,05,576	38,84,371	12.4
17.	Jammu and Kashmir	1,25,41,302	14,93,299	11.9
18.	Maharashtra	11,23,74,333	1,05,10,213	9.4
19.	Andaman and Nicobar Islands	3,80,581	28,530	7.5
20.	Andhra Pradesh	8,45,80,777	59,18,073	7.0
21.	Himachal Pradesh	68,64,602	3,92,126	5.7
22.	Uttarakhand	1,00,86,292	2,91,903	2.9
23.	Kerala	3,34,06,061	4,84,839	1.5
24.	Bihar	10,40,99,452	13,36,573	1.3
25.	Tamil Nadu	7,21,47,030	7,94,697	1.1
26.	Uttar Pradesh	19,98,12,341	11,34,273	0.6

Source: Census of India, 2011

Table 6: Tribal population – lowest and highest

S.N.	Population	State/Union Territory/District	Percentage
1.	State with highest proportion of STs	Mizoram	94.4
2.	State with lowest proportion of STs	Goa	0.04
3.	Union Territory with highest proportion of STs	Lakshadweep	94.5
4.	Union Territory with lowest proportion of STs	Andaman and Nicobar Islands	8.3
5.	District with highest proportion of STs	Sarchhip, Mizoram	98.1
6.	District with lowest proportion of STs	Hathras, Uttar Pradesh	0.01

Source: Census of India, 2011

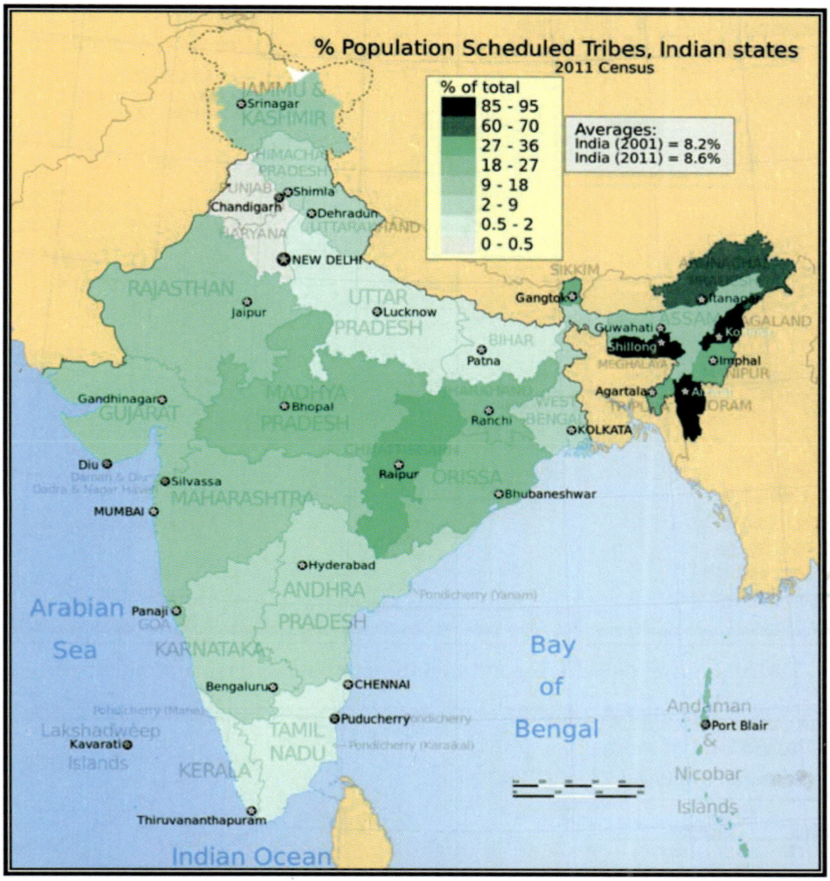

Fig. 24. Percentage population of ST in India (Source: Census of India, GOI, 2011)

4.3. TYPES OF TRIBAL REGION OR GEOGRAPHICAL DISTRIBUTION OF TRIBE

The tribes live all over the country from the foothills of Himalayas to the land tip of Lakshadweep and from the plains of Gujarat to the hills of North-east. There are the tribals in Chamba and Mahasu, and in Dehradun and Darjeeling. The peninsular region has *Bhils*, *Gonds*, *Santhals*, *Ho*, *Kols*, *Kondh*, *Kharia* and *Mundas* among the prominent. In the Northeastern region, the prominent tribals are *Nagas*, *Mizos*, *Mikris* and *Kukis*.

Dr. B.S. Guha has divided the country into following tribal zones:

4.3.1. North–Northeastern Zone

Eastern Kashmir, Eastern Punjab, Himachal Pradesh, Northern Uttar Pradesh, Nagaland, Assam, etc. constitute this zone. *Aka*, *Mishmi*, *Chulikata* and *Naga* are some of the tribes of this zone.

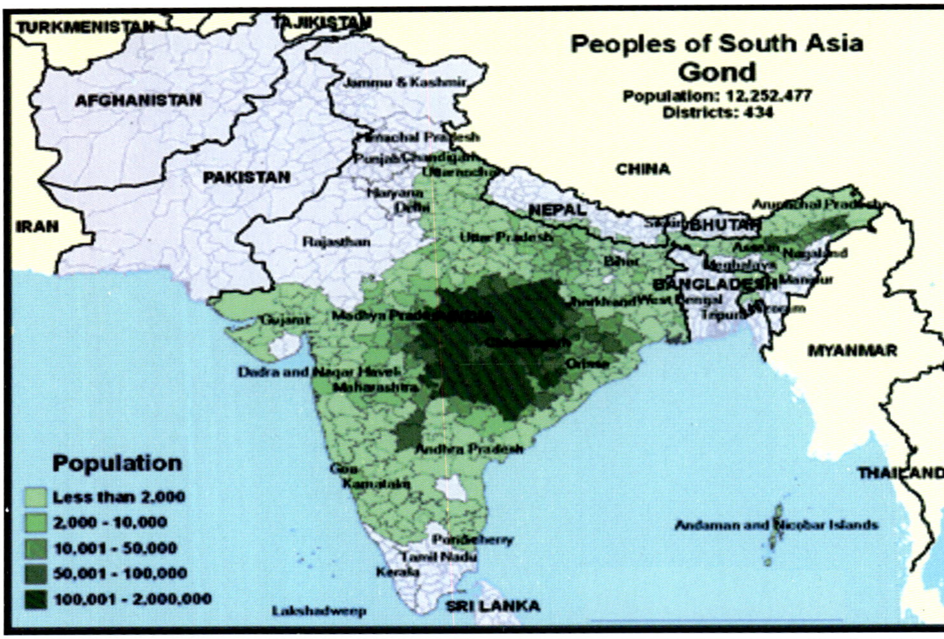

Fig. 25. Distribution of *Gond* Tribe in India (Source: https://joshuaproject.net/)

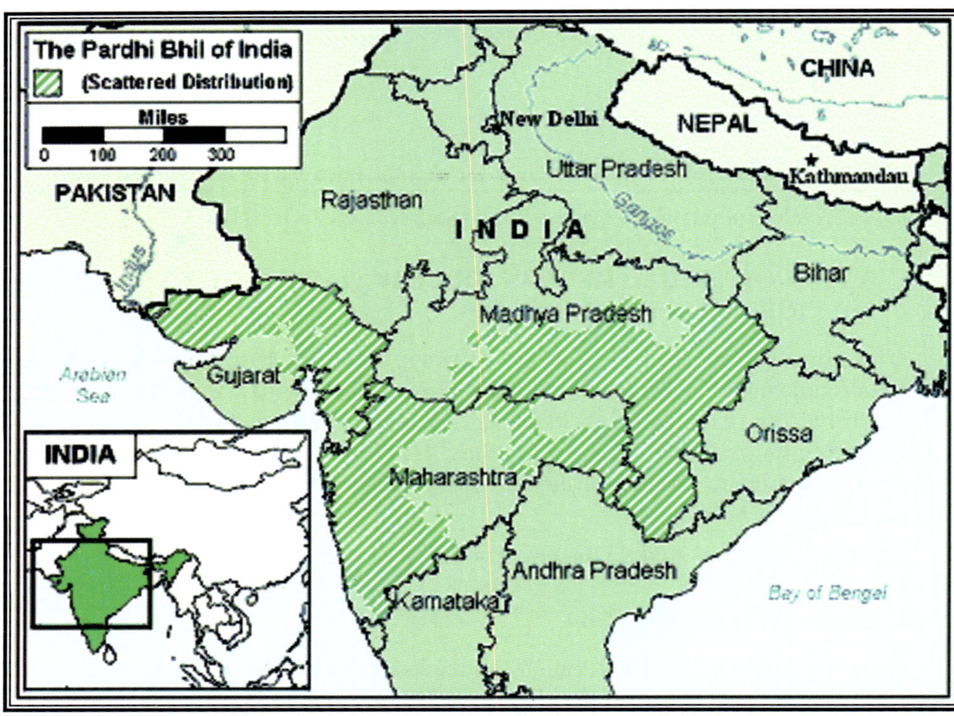

Fig. 26. Distribution of *Pardhi* Tribe in India (Source: https://joshuaproject.net/)

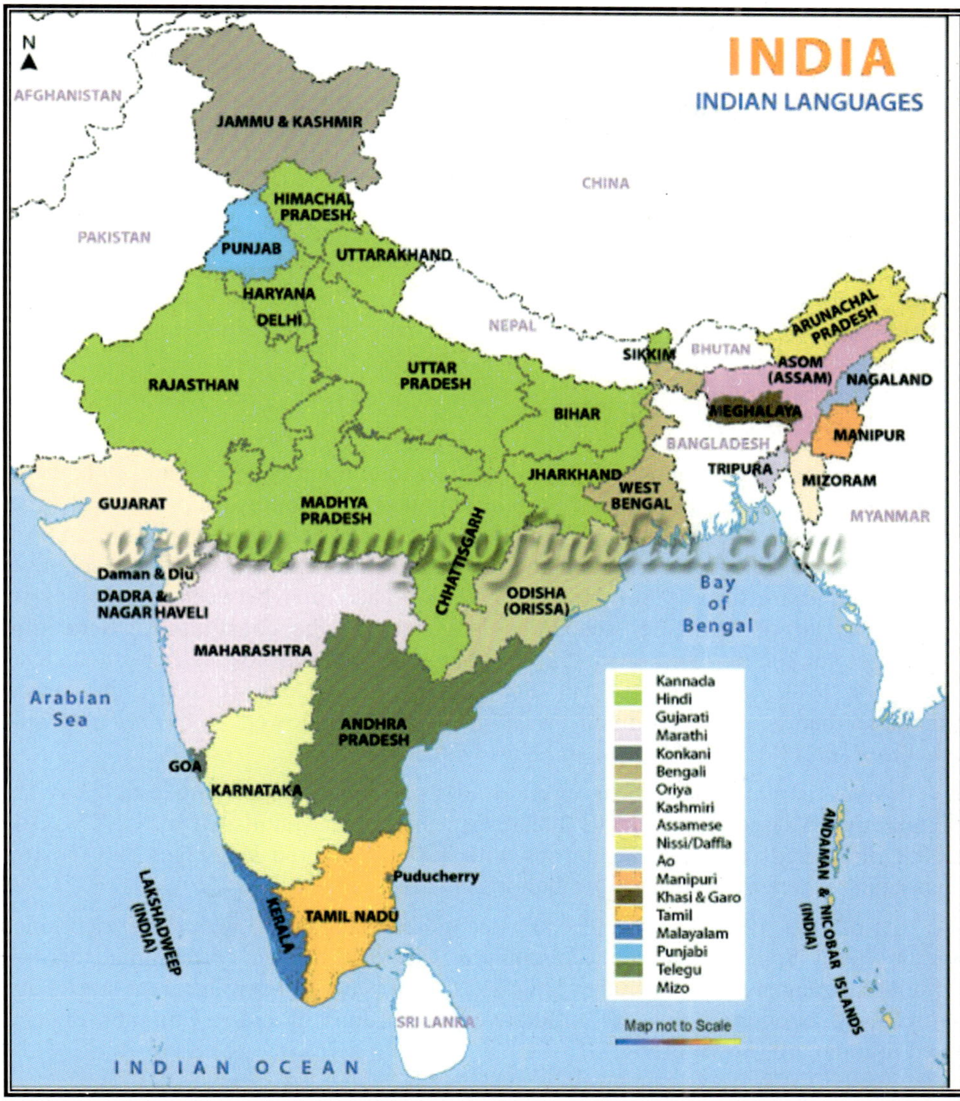

Fig. 27. Common dialects in India (Source: Maps of India, 2011)

4.3.2. Central Zone

Bihar, Bengal, Chhattisgarh, Maharashtra, Southern Uttar Pradesh, Southern Rajasthan, Madhya Pradesh and Odisha come under this zone. *Kondh*, *Santhal*, *Bhil*, *Gond*, *Muria* and *Baiga* are largely populated tribes of this zone. The bulk of the tribal population lives in this zone. There is a very high concentration of tribal population in Central India. Over 85% of the total tribal population inhabits the eight States that constitute this zone.

4.3.3. Southern Zone

The Southern zone consists of four Southern States – Andhra Pradesh, Tamil Nadu, Karnataka and Kerala. *Toda*, *Chenchu*, *Kadar* and *Koraga* are some of the important tribes of this zone.

4.3.4. Andaman and Nicobar Zone

Apart from the other zones, there is a small isolated region comprising Andaman and Nicobar Island in the Bay of Bengal. The principal tribes inhabiting in these regions are *Jarawas*, *Onges*, *Sentinelese*, *Andamanese* and *Nicobaries*. Numerically-smallest tribal community in India is the *Andamanese*, with a population of 19 only.

4.4. TRIBAL DIALECTS

Indian subcontinent consists of considerable number of separate linguistic communities, at times sharing a common language and culture, and again, at times standing in huge difference in dialects. The number of languages listed for India is 418. Of those, 407 are living languages and 11 are extinct. Thirteen languages account for more than 1% of Indian population each, and between themselves for over 95%; all of them are 'scheduled languages of the constitution'. Scheduled languages spoken by fewer than 1% of Indians are *Santali* (0.64%), *Nepali* (0.28%), *Sindhi* (0.25%), *Manipuri* (0.14%), *Bodo* (0.13%) and *Dogri* (0.01%). The largest language that is not 'scheduled' is *Bhili* (0.95%), followed by *Gondi* (0.27%), *Kumaoni* (0.21%), *Tulu* (0.17%) and *Kurukh* (0.10%).

Tribal communities speak around 304 mother tongues. According to the 'Interactive Atlas of the World's Languages in Danger' published by UNESCO in 2009, India leads the list of countries with 196 endangered languages and most of these endangered dialects have a tribal origin.

At present, India has 18 officially-recognized languages, which are mentioned in the Eight Schedule of the Constitution. Regional, *Hindi* and English are the official languages of India. Indian languages have evolved from different stocks and are closely associated with the different ethnic groups of India. Common dialects used in India are shown in Fig. 27.

4.5. LINGUISTIC GROUPS

Broadly, the Indian languages can be put into six groups: 1) Indo–Aryan, 2) Dravidian, 3) Sino–Tibetan, 4) Negroid, 5) Austric, and 6) Others.

4.5.1. Indo–Aryan

This is the most important family of Indian dialects and comprises of all the principal languages of Northern and Western India. It is the part of the Indo–European family of languages, which came to India with the Aryans. It is the biggest of the language groups in India and accounts for about 74% of the total Indian population.

The prominent dialects of this group are: Assamese, *Baigani, Banjari,* Bengali, *Bhatri, Bhili, Bhunjia, Chakma, Chhattisgarhi, Dhanki, Dhodia, Dhundhari, Gadiali, Gamit/Gavti, Garasia/Girasia, Gojri/Gujjari,* Gujarati language, *Hajong, Halbi, Harauti,* Hindi, *Jaunsari, Kachchi, Khotta, Kinnui, Kokni,* Konkani language, *Kotwalia, Kudamamali, Thar, Lambani* or Lamani, *Laria, Magahi, Mahl,* Marathi, *Mavchi, Mewnri, Nagpuri, Naikadi, Nimari,* Oriya language, *Rathi, Sarhodi, Shina, Tharu, Wagri, Warli,* etc.

4.5.2. Dravidian

This is the second most important group and comprises mainly of dialects spoken in the Southern India. This group of languages does not have any relationship with the language groups outside the Indian subcontinent. The Dravidian language came into India centuries before the Indo–Aryan. It covers about 25% of the Indian population. According to the Russian linguist M.S. Andronov, Proto–Dravidian is broadly classified into three groups: Northern group, Central group and Southern group.

The Northern group consists of three dialects i.e., *Brahui, Malto* and *Kudukh. Brahui* is spoken in Baluchistan, *Malto* spoken in Bengal and Odisha, while *Kurukh* is spoken in Bengal, Odisha, Bihar and Madhya Pradesh. The Central group consists of 11 languages viz., *Gondi, Khond, Kui, Dhurwa, Kurukh, Manda, Malta, Parji, Gadaba, Kolami, Pengo, Naiki, Kuvi* and *Telugu.* Out of these, only *Telugu* became a civilized language and the rest remained tribal languages. The Southern group consists of seven dialects viz., *Kannada, Tamil, Malayalam, Tulu, Kodagu, Toda* and *Kota.*

4.5.3. Sino–Tibetan

The *Sino–Tibetan* or *Mongoloid* speech family has a considerably vast expanse in India and stretches all over the sub–Himalayan tracts, covering North Bihar, North Bengal, Assam up to the Northeastern frontiers of the country. The Tibeto–Burman dialects are divided into four broad groups viz., Tibetan, Himalayan, North–Assam and Assam–Burmese.

The important dialects of this region are: *Sikkimese, Bhotia, Balti, Sherpa, Lahuli, Ladakhi, Limbu, Kanauri, Abor* (Adi), *Miri, Aka, Dafla, Mishmi, Kuki–Chin, Mikir, Bodo* and *Naga, Bodo, Rajbangsi, Koch, Mech, Rabha, Dimasa, Kachari, Chutiya, Garo, Haijong* and the *Tipra, Tirupuri, Angami, Sema, Ao, Lotha, Mao, Konyak, Kabui and Lepcha.*

4.5.4 Negroid

The earliest occupants of India were the people of Negrito race and these people migrated to India from Africa and, established their language on the soil of India. The Negrito features are met particularly amongst the Andaman islanders and most probably the *Uralis* of Nilgiri hills, *Kadors* of Kochi, *Pullayans* of Palni hills, etc. Besides some tribes such as the *Angami Nagas* in the north-east and the *Badgis* in the Rajmahal hills in Jharkhand also possess some physical traits reminiscent of the Negrito.

4.5.5. Austric

The *Austric* dialects of India belong to the Austro–Asiatic sub-family, which are represented by languages of the *Munda* or *Kol* group, spoken in the Central, Eastern and Northeastern India, and languages of the *Mon–Khmer* group such as *Khasi* and *Nicobarese*. These are very ancient languages, which have been in existence much before the advent of Aryans and were referred in ancient Sanskrit literature as *Nisadas*. The most important language of the Austric group is *Santhali*, which is spoken by over five million *Santhals* and is the largest spoken among the *Adivasi* languages. *Mundari*, spoken by about a million *Mundas*, is another important language of this group.

4.5.6. Others

There are several Dravidian *Adivasi* dialects such as *Gondi, Oraon* or *Kurukh, Mal–Pahariya, Khond* and *Parji,* which are very distinct and cannot be classified in other groups.

Note:

- Detailed list of important Indian tribes with their major characteristics are given in **Annexure I (Table 20)**.
- The list of Notified Indian STs is shown in **Annexure II (Table 21)**.
- Detailed list of tribal dialects commonly used by Indian tribes is given in **Annexure III (Table 22)**.

5

TRIBAL ECONOMY: ECONOMIC TYPOLOGY AND CHARACTERISTICS

5.1. INTRODUCTION

The primitive societies have passed through several stages of economic development everywhere in the world. There are four important stages, which have been identified in the economic development of tribals, these are: hunting and food-gathering stage, pastoral and pre-agricultural stage, agricultural stage, and post-agricultural and technological stage.

5.2. ECONOMIC TYPOLOGY OF THE TRIBE

The sources of subsistence and livelihood are varied so far as the Indian tribals are concerned. Starting from the pure and simple parasitic habit of the nomadic hunters and food-gatherers that depend mostly on nature for the sources of subsistence to the settled agriculturists and the group of industrial labourers, we have the views of different economic set-up of the Indian tribals. The economic typology of the Indian tribes as proposed by Vidyarthi (1985) into eight types is given below:

5.2.1. Forest Hunting Types

The tribe falling in this economic typology, principally, depends on food gathering and hunting for their livelihood. This type of tribe is distributed all over the country. The tribes living in dense forests full of wild beasts live on hunting.

This mode of livelihood of the earliest times still continues in a few tribes who usually depend on the existence of some surpassingly abundant wild roots, fruits and other palatable food e.g., leaves and flowers. These forest-hunting type tribes have mainly three resources of economy, which vary according to season, cycle and area – (a) food gathering, (b) hunting, and (c) fishing. The hunters leave the females to carry out household activities in the morning and return in the evening after hunting. In some tribes, there is a usual custom of hunting collectively. The *Nagas* use spears and arrows and bows. The *Bhils* are very much specialized in hunting by arrow. Some of the hunting type tribes are *Rajis*, *Birhor*, *Kadar*, *Muria*, *Pardhi*, *Kharia*, *Korawa*, etc.

The nomadic primitive tribes such as the *Andamanese*, *Onge*, *Jarwa*, *Kadar*, *Kharia*, and *Lodha* are included in this category so far as their subsistence economy is concerned. They usually live far away from rural-urban way of life and possess a simple type of social organization.

5.2.2. Hill Cultivating Types

This type mainly practices shifting cultivations. They adopt skilled methods and techniques for this purpose and it is common in North-eastern States of India. A section of the *Gond*, *Naga*, *Kharia*, *Juang*, *Mogs*, *Nagas*, *Santhal*, *Riang*, *Khasi*, *Garo*, and *Savara* practice such type of primitive cultivation by 'slash and burn' method. It is known by different names in different areas. The Assam tribes call it *Jhum*, the *Gond* as *Podu*, etc. A hilly-forested tract selected for this purpose may have to be abandoned after three successive cultivating seasons as this soil is likely to lose fertility. The plants, shrubs, and undergrowths are cut down, and left for drying for a month or so. Then, they set fire to them. The ashes serve as auto-manure to the soil. On the onset of monsoon, the soil is slightly loosened by a simple digging stick or hoe. Seeds of different Kharif crops, millets such as bajra, jowar, kurthi, pulses, potato, paddy, and tobacco are grown in this type of cultivation. It can support them partially, but not wholly. They have to find out some other vocations as supporting source of subsistence.

5.2.3. Plain Agriculture Types

Most of the tribes of India practice agriculture as a major economic activity, but this agriculture is simple in nature and the implements used are crude. The tribals find it safer than hill cultivation and much safer than forest hunting. A large percentage of the tribals are supported by plain agriculture. All the major tribes of India practice agriculture as their primary source of livelihood.

Like the cultivators of the advanced societies, some of the tribals in India are taking resort to permanent settled agriculture. The *Oraon*, *Khasis*, *Jaintias*, *Bhill*, *Munda*, *Gond*, *Bhumij*, *Ho*, *Santal* are efficient cultivators at present. They practice wet cultivation by transplanting method. Artificial irrigation and application of compost manure are unknown to them. Rotation of crops is within the knowledge of these cultivators. They work in their own fields as well as in the fields of others as share-croppers (Bhag–Chasi). The social and religious organizations of these settled agriculturist tribals are much developed and highly complex.

5.2.4. Simple Artisan Types

Some of the tribals are still retaining their traditional crafts along with the principal sources of subsistence. These tribes derive their livelihood from the sale of articles of art and craft prepared by them or through cottage industry such as weaving cloths, basket making, tool making, metal work and iron smithy. Preparing ropes and skins, and utensils of different metals are prevalent in many tribes. The *Kharia* people are very much specialized in cottage industries. Handicrafts are a subsidiary occupation for most of the advanced tribals such as *Gujjar*, *Kanjars*, *Kotwalis* and *Mahalis*. The *Naga* and the *Khasi* are experts in coloured-handloom products. With the marginal profit in their traditional specialized crafts, these tribals are, at present, taking resort to other types of jobs. Their mixed pattern of economy reflects back adversely upon their social system.

5.2.5. Pastoral and Cattle Herder's Type

The main occupation of these tribes is cattle rearing. The hilly tribes rear the cattle, examples of which are *Gujjars* and the tribes of Chamba, the *Bhotia* of Almora and *Todas* of Nilgiri hills of South India. *Bakarwalas* and *Gaddis* are also a purely pastoral tribe. They do not practice agriculture, hunting, fishing, etc. They also live far away from the sophisticated modern world and possess a slightly developed, but non-complicated social structure. They rear buffaloes, cows, and goat; the milk-products are being exchanged to procure the commodities of day-to-day use.

5.2.6. Folk Artist Types

Tribes falling under this typology are those, which have adopted bard singing, singing, dancing, music, conjuring, acrobatics, snake charming, etc., as their means of livelihood. Actually, these people are folk artists and economically dependent on their art. The Indian tribes that can be captioned as folk artists are the *Nats* and *Saperas* of southeastern Uttar Pradesh, the acrobat *Mundupttas* and *Kelas* of Odisha, the bard singing *Pardhans* and *Ojhas*, the animal players *Madarias*, the snake charmer *Pamulas*, the *Garadis*, the acrobat *Dommaras* and *Bynollus*, the conjuring *Palukumuggulas* and *Paddintigollas*, the *Bahuroopas*, etc. of Andhra Pradesh, a few *Kotas* of Tamil Nadu and the snake charmer *Kalbelias* of Rajasthan.

5.2.7. Agriculture and Non-agriculture Types

Tribals work in agricultural or non-agricultural industries on a casual basis. They may get employment on a daily basis. Agricultural work possibilities exist in the local, nearby villages and also in different States. Tribals from this group go out as seasonal migrants to near by or distant tours in mines, mills and tea gardens to work as labourers, in railway and road construction, forestry, construction such as civil work, houses, dams, bridges and bricks unit.

5.2.8. Skilled White Collar Job, Traders Types

Some individuals or families of tribal communities of all the regions are working for their livelihood in different offices, hospitals, factories and business enterprises. Some of them are engaged in commercial trade also. Apart from their initiation, the seat reservation in different services as laid down in the Constitution has helped a lot in creating this type of economic life among the tribals.

5.3. STRUCTURE / GENERAL CHARACTERISTICS OF TRIBAL ECONOMY

The tribal economic organization is mainly concerned with producing such things, which are necessary for their daily needs, and consuming them. These activities are very much determined by the geographical environment of the tribe.

Usually, the tribals have to struggle very hard to meet their economic needs. The following are the important characteristics of Indian tribal economic organization:

5.3.1. Forest-based Economy

The most important character of the tribal economy is the close relationship between their economic life and their natural environment and habitat which is the forest. This economy revolves around the forest. Tribals obtain their numerous requirements from the area they inhabit with the help of most simple implements and without technological aid from outside. The edible fruits, flowers, roots, bark, seeds, vegetables, honey, insects, meat of wild animals, fish, and other flora and fauna are collected from the forest for their livelihood. The *Bhill* of Western India depend on mahua flower and bidi leaves, the basket makers *Mahalis* of Bihar get bamboo from forest area, the pastoral *Gaddis* of Himachal Pradesh depend on forest for the fodder for their animals.

5.3.2. Division of Labour

All the members of the family together form a production unit. Generally, wives do most of the work such as cooking of food, household work, collection of minor forest produce, firewood and fuelwood from forest, helps in agriculture work such as sowing, weeding and harvesting. Children of the tribals engage in grazing of animals and also help their mothers in different household work. Husband takes part in preparation of land, ploughing, harrowing, watch and ward, harvesting of the crops, collection of honey, fishing, hunting, etc. Old women generally stay at home. The whole family works for their livelihood. There is a well-organized pattern of division of labour in a family, which is based on age and sex alone. Male, female and children can work according to their physical capacity.

5.3.3. Simple Technology

The tribals use indigenous implements for hunting and agriculture purpose. Use of such crude implements involves a considerable amount of time, manual labour and waste of energy. The tribals, in general, do not resort to the use of technological aids. As the tribals are illiterate and cut off from the civilized world, they generally carry out production without adequate technological aids with the result that there is much loss of material with very little production. The tribals are, therefore, generally very poor in spite of working very hard.

5.3.4. Absence of Profit in Economic Dealings

In economic dealings, the element of profit is quite absent among the tribals. The barter system fulfils ones requirements, but does not give any profit. The economic dealings among the tribals are mostly based on exchange.

5.3.5. Tribal Community Acts as a Co-operative Unit

Natural co-operation within the tribe constitutes one of the striking features of the pattern of labour. Many types of works are done by them collectively or in a group e. g., common grazing of animals, working in the shifting cultivation and settled agriculture fields, joint activity for watch and ward of fields, food gathering such as honey, some minor forest produce, hunting of animals, fishing, etc.

5.3.6. Gift and Ceremonial Exchange

The exchange exists as a part of an economic relationship among the tribals. Gift giving and hospitality to social intimates is an important part of their economy. In the villages, local artisans such as ironsmith, wood carpenter receive cash or kind for a year for repair of agriculture implements.

5.3.7. Periodical Market

There are no permanent markets existing in the tribal community, only periodical weekly or fortnightly market. The tribal people do not use currency in deciding the price of the commodities; the economic transaction takes place in the form of barter for native goods. The non-native goods have to be purchased in cash.

5.3.8. Economic Backwardness

The above mentioned characteristics of tribal economic organization account for their economic backwardness. They do not know the new changes and inventions in different fields of production. Their methods and implements of cultivation, hunting and fishing are very primitive. They do not know anything about trade and commerce. The cottage industries are carried on by means of unrefined and primitive tools and methods. Hence, the economic status of Indian tribes is very much backward.

6

TRIBAL: SOCIO-ECONOMIC AND CULTURAL PROBLEMS

6.1. INTRODUCTION

Tribal communities continue to be vulnerable even today, not because they are poor, assetless and illiterate compared to the general population; but often their distinct vulnerability arises from their inability to negotiate and cope with the consequences of their integration with the mainstream economy, society, cultural and political systems, from all of which they were historically protected by their relative isolation. The requirements of planned development brought with them the dams, mines, industries and roads, all located on tribal lands. With these, came the concomitant processes of displacement followed by a conflict between development and protection of tribal rights and interests. Tribal institutions and practices were forced into uneasy co-existence, which paved the way to market or formal State institutions. Also, the tribals found themselves at a great disadvantage in the face of an influx of better-equipped outsiders into tribal areas. The repercussions for the already fragile socio-economic sustenance base of the tribals were devastating – ranging from the loss of livelihoods and land alienation on a vast scale to hereditary bondage.

There are many acute problems of the tribal people in our country that need immediate attention and early solution. The problems relate to various aspects of tribal people *viz.,* social, economical, educational, health, religion, land, law and order situation, self-centered tendency and so on. Many of the problems cannot be well-understood due to lack of necessary and adequate information. Tribal cultures have ancient roots, the lack of infrastructural development and at times inaccessibility of remote areas leads to a number of problems. The major problems and / various forms of exploitations of the tribals are discussed below:

6.2. SOCIO-ECONOMIC AND CULTURAL PROBLEMS OF TRIBALS

6.2.1. Problems Related to Forests

In the forests, tribes lived, enjoyed hunting, gathering and shifting of cultivation and articles made from forest material. This was the only source for their livelihood. The backbone of forest dwellers' economy is the flora and fauna found in the nearby forests. But forest policies and regulations have snatched away traditional rights of STs over forest. The various kinds of restrictions imposed on the forest dwellers virtually put them at the mercy of the forest department, especially lower-level functionaries. Illiteracy and poor economic conditions make them even more vulnerable.

Due to this, economic activities of tribals such as food gathering, hunting, collection of non-timber forest produce, shifting cultivation and cottage industry have been affected adversely. Forest land and its resources provide the best means of livelihood for the tribal people and many tribes, including the women, engaged in agriculture, food gathering and hunting. They are heavily dependent on the products of the forest. Therefore, when it adversely affects the tribe's life and its resources, the natural life cycle of tribal ecology and tribal life is greatly disturbed.

6.2.2. Indebtedness

Majority of the tribal community is chronically in debt. They take loan for marriage, birth, death, social and religious purposes. Most tribals find it difficult to repay the loan. Once a family takes loan from the local money lender, it is very difficult to pay it back and the debt of one generation is passed on to the other generation. Debt descended from father to son even passes to the next generation. The tribal becomes easy victim for the moneylender, contractors, land lord, and market middlemen.

The problem of indebtedness amongst tribals is not only an indication of their poverty, but it also reflects the wider economic malaise i.e., lack of education, low purchasing/bargaining power and lack of resources to engage in a gainful activity and meeting emergent expenditure. Therefore, the problem continues to persist with increasing menace as the indebtedness further pushes the tribals into extreme conditions of poverty. The indebtedness forces them to dispense with their meager resources including the small bits and pieces of land to pay off the loans at exorbitant rates of interest. The initiation of commercial vending of liquor in tribal areas has started impoverishing the tribal population, making them victims of indebtedness and exploitation.

6.2.3. Land Alienation

According to latest statistics, nearly 88% of the STs are engaged in agriculture. The tribals have great emotional attachment with their lands. Agricultural is the only source of livelihood, which most of them have known for centuries.

The socio-economic survey of tribals reveals that they alienate their lands mainly for meeting their domestic expenses and for clearing their past debts. The main causes of land alienation and landlessness among the tribals are poor economic conditions, need of cash for various purposes such as marriages, birth, death, fairs and festivals, drinking habit, indebtedness, chronic shortage of cash, poor and inadequate agricultural production, urbanization, industrialization, lack of land records, administrative inefficiency, delay in getting judgement, oral mortgage of lands in the hands of non-tribals, non-possession of judgement, and interest not shown by tribal pleaders or not taking interest due to heavy bribe, fear of police.

Land alienation has given birth to several allied problems among the tribals i.e., increased poverty among the tribals, decreased opportunities of employment, migration of tribals, exploitation of tribal labourers, exploitation of tribal women, created tension between tribals and non-tribals, increased distance between the rich and the poor tribals, developed extremism and naxalism in tribal areas.

Serious land-alienation problems have come up in the tribal areas. A number of existing laws prohibit the transfer of land from tribals to non-tribals. Marriage of tribal girls with non-tribals also resulted in land alienation. The tribal lands are occupied by money lenders, *Zamindars* and traders by advancing loans to the tribals. Rich tribals are also purchasing the land of poor tribals. Government acquired the tribal lands for various developmental activities such as mines, power projects and protected area. In this way, land alienation has resulted in landlessness among the tribals. Now, all transfers of tribal lands are banned unless permitted by an appropriate authority.

Korwa tribe of Sarguja district of Madhya Pradesh, and *Tharu* and *Bhoksa* of Tarai of Uttar Pradesh have been the worst victim of deceit and plunder of their land. The entire region inhabited by the *Korwas* is surrounded by money lenders locally known as *Miyan*. They are ever ready to advance loan at the slightest pretext with the only condition that the *Korwa* has to mortgage some of his land. The *Tharu* and *Bhoksa* tribes of Tarai region of Uttar Pradesh are the victims of Punjabi refugee money lenders and traders who have grabbed thousands of acres of land reclaimed by the tribals.

6.2.4. Bondage Labour

The STs belonging to Odisha, Chhattisgarh, Haryana, Madhya Pradesh, southern Uttar Pradesh, Uttaranchal, Andhra Pradesh, Maharashtra, Rajasthan and Gujarat who have suffered a gradual erosion of access to traditional livelihood systems, have long been subjected to exploitative debt relations leading to loss of land and bondage to non-tribals. The practice of bondage labour is called *Ajalu* in Karnataka, *Palamuuru* in Andhra Pradesh and *Wetbighari* in Maharashtra State.

Labour bondage has been reported in industries/works of stone quarries, crushers and mines, brick kilns, fish processing, silver work, mat weaving, powerloom and cotton handloom, construction, plantation, gem cutting, rice mill, bidi industry, salt pan workers, agriculture, etc. In some villages, landlord or *Zamindar* gives fellow land to the tribals for cultivation and take more part of income or paid money to the tribal in advance at the time of their needs and bondage them as slavery and engaged them as a labour.

6.2.5. Middleman and Traders

Market facilities are not available in tribal areas for selling their agricultural and forest produce hence, middleman and traders supply essential commodities such as salt and match box with exchange of agriculture or forest produce. Tribals do not have knowledge and information on market rate and correct weight for their forest collection. Hence, the traders cheat them every time.

6.2.6. Contractor

In tribal areas, the main occupation of the tribals is collection of forest produce, cutting of timber, etc. The contractor takes contract of the harvesting of timber, road construction and other works. The contractor pays fewer wage and gets more work hours in a day.

6.2.7. Chronic Poverty

Majority of the tribes live under poverty line. The tribes follow many simple occupations based on simple technology. Most of the occupation falls under the primary occupations such as hunting, gathering, and agriculture. The technology they use for these purposes belongs to the most primitive kind. There is no profit or surplus in such an economy. Hence, their per capita income is very meagre; much less than the Indian average. Most of them live under abject poverty and are in debt in the hands of local moneylenders and *Zamindars*. They often mortgage or sell their land to the moneylenders to repay the debt. Indebtedness is almost inevitable since heavy interest is to be paid to these moneylenders.

Table 7: Statewise percentage of ST population below poverty line (2009-10)

Sr. No.	State	Rural	Urban
1.	Andhra Pradesh	40.2	21.2
2.	Assam	32.0	29.2
3.	Bihar	64.4	16.5
4.	Chhattisgarh	66.8	28.6
5.	Gujarat	48.6	32.2
6.	Himachal Pradesh	22.0	19.6
7.	Jammu and Kashmir	3.1	15.0
8.	Jharkhand	51.5	49.5
9.	Karnataka	21.3	35.6
10.	Kerala	24.4	5.0
11.	Madhya Pradesh	61.9	41.6
12.	Maharashtra	51.7	32.4
13.	Odisha	66.0	34.1
14.	Punjab	16.1	15.0
15.	Rajasthan	35.9	28.9
16.	Tamil Nadu	11.5	17.6
17.	Uttar Pradesh	49.8	20.2
18.	Uttarakhand	20.0	0.0
19.	West Bengal	32.9	20.6
All India		47.4	30.4

Source: Planning Commission Report, GOI (2014)

Proportion of STs below the poverty line is substantially higher than the national average. As per the statement provided by the Planning Commission, it is observed that ST people living below the poverty line in 2009-10 were 47.4% in the rural areas and 30.4% in the urban areas respectively. It clearly indicates the higher level of poverty amongst the STs. Statewise details regarding poverty among the ST is shown in Table 7.

6.2.8. Migration for Employment

The tribals have to migrate to different places in or out of the State, generally for work and wages, during off season. Metropolitan cities such as Delhi, Chennai, Mumbai, and Kolkata are their major destinations for searching some gainful employment/casual labour in the unorganized sector and as household maids for their livelihood. They are often exploited physically, economically and sexually. Generally, tribal migrants have find jobs in civil construction work, canal digging, road construction, bricks kiln, small factories, agro-processing plants or working as porters, domestic servants, bus cleaners, rickshaw pullers, street hawkers, petty traders, workers and domestic workers.

According to the 2011 census figures, 34.5% of the ST population were cultivators, 44.5% were agricultural labourers, 1.8% were household industry workers and 19.2% were other occupation workers.

6.2.9. Health and Malnutrition

In many parts of India, tribal population suffers from chronic infections and diseases out of which waterborne diseases are life threatening. They also suffer from deficiency diseases. The Himalayan tribes suffer from goiter due to lack of iodine. Leprosy and tuberculosis are also common among them. Infant mortality was found to be very high among some of the tribes.

The nutritional status of various STs varies from tribe to tribe, depending upon the social, economic, cultural and ecological background. Though no systematic and comprehensive research investigations have been carried out, it appears that malnutrition amongst the tribe, especially tribal children and women, is fairly common. It debilitates their physical condition, lowers resistance to disease, and in the post-weaning period, leads, at times, to permanent brain impairment. To quote the Ninth Plan Working Group on the Tribal Development, Experts have opined that not a single tribe in the different States of India can said to be having a satisfactory dietary pattern as tribal diets are frequently deficient in calcium, vitamin A, vitamin C, Riboflavin and animal protein.

Malnutrition is common and has affected the general health of the tribal children as it lowers the ability to resist infection, leads to chronic illness and sometimes leads to brain impairment. Proper medical health facilities are not available in tribal areas. Further, high incidence of malnutrition is observed, especially among primitive tribal groups in Phulbani, Koraput and Sundergarh districts of Odisha as also amongst the *Bhils* and *Garasias* of Rajasthan, the *Padhars*, *Rabris* and *Charans* of Gujarat, *Onges* and *Jarawas* of Andaman and Nicobar Islands and *Yerukulas* of Andhra Pradesh, etc.

The Infant Mortality Rate (IMR), Under-5 child Mortality Rate (U5MR) for STs as well as of other disadvantaged socio-economic groups for some of the States are as shown in the Table 8. Percentage difference in IMR for ST population is very high in Andhra Pradesh (25.7), Gujarat (23.2) and Manipur (21.5), whereas, for the States of Assam, Jammu and Kashmir, Karnataka and Sikkim, the IMR for ST population is lower than the general population by 7.1, 11.2, 7.2 and 4.8 percentage points respectively. Likewise, percentage difference in the U5MR for ST population is very high in Odisha (41.6), Gujarat (38.8), Andhra Pradesh (33.3), and Madhya Pradesh (32.5). The U5MR for ST population for the States of Sikkim and Assam is lower than the general population by 4.2 and 1.8 percentage points respectively.

Table 8: Statewise child health indicators

Sr. No.	State	IMR		U5MR	
		All	ST	All	ST
1.	Andhra Pradesh	68.4	(94.1)	78.7	(112)
2.	Arunachal Pradesh	60.7	67.6	87.7	100.9
3.	Assam	66.1	59	85	83.2
4.	Chhattisgarh*	80.8	90.6	105.5	128.5
5.	Gujarat	62.8	(86.0)	77.0	(115.8)
6.	Jammu and Kashmir	45.5	(34.3)	53.8	*
7.	Jharkhand*	76.6	93.0	112.4	138.5
8.	Karnataka	53.0	(45.8)	66.2	(77.9)
9.	Madhya Pradesh	81.9	95.6	108.2	140.7
10.	Maharashtra	45.3	51.4	53.4	69.8
11.	Manipur	29.7	51.2	41.9	71.4
12.	Meghalaya	44.6	49.3	70.5	74
13.	Nagaland	38.3	45.8	64.7	65.8
14.	Odisha	67.7	78.7	94.7	136.3
15.	Rajasthan	72.7	73.2	93.3	113.8
16.	Sikkim	33.7	28.9	40.1	35.9
	All India	57	62.1	74.3	95.7

Source: Annual Report - 2013-15, Ministry of Tribal Affairs (MoTA), GOI

( ) Based on 250-499 unweighted cases

* Rate not shown based on fewer than 250 unweighted cases surviving to the beginning of the age interval.

6.2.10. Education

Illiteracy is a major problem of the tribals. According to Census 2011, only 59% of the tribal population is literate. They have no faith in formal educational organization. Many of them do not know anything about education, schools, colleges, universities, degrees, etc. They feel no urge to educate their children. Since most of the tribals are poor, it is not easy for them to send their children to schools, as they are considered as extra helping hands for agriculture and other household work. The formal schools do not hold any special interest for the children. Most of the tribes are located in the interior and remote areas where schooling facilities are not available.

Statewise Literacy Status among ST

Among the States, Mizoram and Lakshadweep have highest Literacy Rate (LR) for STs, while the lowest LR was observed in Andhra Pradesh and Madhya Pradesh. Ten States having LRs less than the country's average (58.96%) for ST population in census 2011 are indicated in Table 9.

Table 9: States having ST LRs less than country's average for STs

Sr. No.	State/Union Territory	LR
	India	**58.96**
1.	Andhra Pradesh	49.2
2.	Madhya Pradesh	50.6
3.	Jammu and Kashmir	50. 6
4.	Bihar	51.1
5.	Odisha	52.2
6.	Rajasthan	52.8
7.	Tamil Nadu	54.3
8.	Uttar Pradesh	55.7
9.	Jharkhand	57.1
10.	West Bengal	57.9

Source: Census of India, 2011

6.2.11. Lack of Communication and Transportation Facilities

The tribals live in remote areas. Hence, one has to face a great problem of communication. It takes much time for news related to welfare and development to reach them. As a result, they are not conscious of their rights and are ignorant of a number of development schemes.

6.2.12. Displacement of Tribals

Displacement or forced/voluntary eviction of tribals from their land and their natural habitats, and subsequent rehabilitation has been a serious problem that remains to be addressed by the Government. As per the readily available information, a population of 21.3 million has been displaced between 1951 and 1990 in the States of Andhra Pradesh, Bihar, Gujarat, Maharashtra, Madhya Pradesh, Rajasthan and

Odisha. Of them, 8.54 million (40%) are tribals and of those only 2.12 million (24.8%) tribals could be resettled, so far. Displacement took place mainly on account of development projects, which include large irrigation dams, hydroelectric projects, opencast and underground coal mines, super-thermal power plants and mineral-based industrial units. In large mining projects, tribals lose their land not only to the project authorities, but even to non-tribal outsiders who converge into these areas, and corner both land and the new economic opportunities in commerce and petty industry. The incomplete rehabilitation of the displaced tribals has further compounded their woes as they are pushed into a vortex of increasing assetlessness, unemployment, debt bondage and destitution. Women and children, as ever, are the worst affected.

6.2.13. Alcoholism

Alcoholism and other pschycological and socio–cultural ramifications of alcoholism constitute the major problem among the tribal people. Brewing of alcohol from rice, millets, 'mahua' flower and fruits had been practiced traditionally. However, in recent years, inroads have been made by businessmen by making inexpensive molasses and ammonium chloride available for making alcoholic drinks for home breweries. Consequently, very large quantities of cheap liquor are brewed. Availability of readymade and much stronger liquor has made the tribal an alcoholic. Alcoholism is, therefore, one of the major problems of the tribal societies.

6.2.14. Cultural Problems

The tribal culture is entirely different from the culture of the civilized people. The tribals fail to understand the civilized people, their customs and practices, beliefs and attitudes. They are suspicious towards the civilized people. They are clinging tenaciously to their customs and traditions.

Due to the contact with other cultures, the tribal culture is undergoing a revolutionary change. Due to an influence of Christian missionaries, the problem of bilingualism has developed, which has led to an indifference towards the tribal language. The tribal people are imitating Western culture in different aspects of their social life and leaving their own culture. It has led to degeneration of tribal life and tribal arts such as dance, music and different types of craft.

6.2.15. Social Problems

The tribals have their own social problems also. They are traditional and custom-bound. They have become the victims of superstitious beliefs, outmoded and meaningless practices and harmful habits. Child marriage, infanticide, homicide, animal sacrifice, exchange of wives, black magic and other harmful practices are still found among them. They believe in ghosts and spirits. They have keen desire to maintain all these practices in general, and their individual tribal character.

6.2.16. Agriculture and Shifting Cultivation

Agriculture and shifting cultivation is still being practiced by the tribal population on higher slopes of hilly areas of the country. As estimated, more than six lakh tribal families in the States of North-east, Odisha, Andhra Pradesh, Himachal

Pradesh, Bihar and Jharkhand practice shifting cultivation, which is ecologically unsound. The shifting cultivation is integrally linked to the tribal economy in the areas where it is practiced and their social, economic and ritual activities are also centered on this practice. The problem of shifting cultivation is a very complex one involving economic, social and psychological aspects of the tribal communities.

The Ministry of Agriculture has been implementing a scheme for control and transformation of shifting cultivation in the Northeastern States, but the pace of its implementation has been very slow. Rehabilitation of shifting cultivators should be given priority for improving ecology and economy of the tribal areas.

6.2.17. Drinking and Irrigation Water

Inadequate water resource is a major problem in many tribal villages and habitations, and particularly those in the interior of hilly areas, are devoid of safe drinking water; and irrigation facilities. However, the National Water Policy adopted in 1987, inter-alia, focuses on the development of water management systems for both drinking purposes and irrigation, based on an integrated approach, to fulfill the needs of the disadvantaged sections of the society.

Water for drinking under the Accelerated Rural Water Supply Programme (ARWSP), especially through the Mission Mode Approach (MMP) adopted in rural and tribal areas, about 10.6 lakh (74.2%) habitations have been covered fully and 3.1 lakh (21.9%) habitations partially so far, with the provision of a safe drinking water source, as against the total 14.3 lakh identified habitations to be covered in the country. This leaves behind only 3.9% of the problem of the villages/habitations. But, unfortunately, no exclusive information is available about the number of tribal habitations covered under the programme. The problem of drinking water is more acute in hamlets located in higher level hills where ground water cannot be tapped, and surface water is difficult and costly to be carried through the scheme to the habitation.

6.2.18. Intellectual Property Rights (IPRs)

The tribals have developed invaluable indigenous knowledge by living closely with the forests and also interacting constantly with the various flora and fauna. This indigenous knowledge amongst the tribal communities is passed on from generation to generation through the medium of oral tradition, folklore and practices, which find a place in various life-cycle events including treatment of diseases/ ailments, without any codified text or rights to accredit their legitimate ownership. As tribal communities have a very close dependence on biological resources, their livelihood and lifestyle often depends upon and are shaped by these resources. Therefore, their survival and sustenance is closely linked to conservation and utilization of these resources. Corporate protectionism in terms of patents and IPRs arising out of various international treaties/instruments on trade and common property resources such as the Trade-related Aspects of Intellectual Property Rights (TRIPS) under the World Trade Organization (WTO) represents a real threat to economic livelihood of these communities. They are also a source of potential exploitation of the tribal resource base as bio-diversity expressed in life forms and knowledge is sought to be converted into private property and treated as an open access system for free exploitation by those who want to privatize and patent it. There is an urgent need to provide appropriate legal and institutional arrangements for recognizing and acknowledging the rights of tribal communities to such resources and knowledge.

6.2.19. Extinction of Primitive Tribal Groups

The Government of India identified 75 tribal communities as Primitive Tribal Groups (PTGs) with a total estimated population of 1.32 million in 1991 spread over 15 States/Union Territories. There is a marked difference between the conditions of the relatively-advanced tribal groups and the PTGs. When compared to the other tribals, they are subjected to extreme backwardness as they live in the most interior and inaccessible forests. A decline in their sustenance base and the resultant food insecurity, malnutrition and ill health force them to live in the most fragile living conditions and some of them are even under the threat of getting extinct. Prominent examples in this context are the Bay-islanders such as the *Great Andamanese, Shompens, Jarawas* and *Sentinelese* of the Andaman and Nicobar Islands. Even some of the mainland groups, which can be cited in this context such as the *Bondos* of Odisha, *Cholanaickans* of Kerala, the *Abujhmarias* of Chhattisgarh and the *Birhors* of Jharkhand, are also dwindling.

6.2.20. Deprivation of Forest Rights

Forests and tribals share a symbiotic relationship. Tribals continue to live in forest areas, though in isolation, yet in harmony with environment. Recognizing this dependency, the National Forest Policy, 1988, stipulated that all agencies responsible for forest management should ensure that the tribal people are closely associated with the regeneration, plantation, development and harvesting of forests so as to provide them gainful employment. Despite these special safeguards, tribals continue to struggle for mere survival as they face formidable problems and displacement due to development of national parks and wildlife sanctuaries and other environmental restoration projects, lack of development in forest villages, etc. The protection of rights of tribals in forests is the key to the amelioration of their conditions.

6.2.21. Crimes Committed against ST

The statistical profile of crimes committed against ST population in India when compared with All-India incidences, indicates that highest percentage of kidnapping and abduction of ST population is noticed in Madhya Pradesh (28.5%), followed by Jharkhand (21.9%) and Rajasthan (13.1%). In cases of rape (Sec. 376 IPC), Madhya Pradesh again scores the highest with 39.6% of ST population linked to such incidents. Kerala is next with14.1% of incidences. The lowest percentage among the States is seen in Karnataka (0.3%) and Tamil Nadu (0.4%). In the cases of murder (Sec. 302 IPC) against ST population, again Madhya Pradesh is highest with 34.3% incidences, Chhattisgarh is next with 15.4% cases and Rajasthan with 13.3% incidences. The least number of cases is seen in Tripura and Bihar with 0.7% incidences. It can, therefore, be surmised that the maximum incidences of crimes against ST population as percentage contribution to All-India is seen in Madhya Pradesh, Jharkhand, Rajasthan and Chhattisgarh (Annual Report, MoTA, 2015).

7

TRIBALS AND THEIR DEVELOPMENT

7.1. INTRODUCTION

The greatest challenge that the Government of India has been facing since independence is the proper provision of social justice to the ST people, by ameliorating their socio-economic conditions. STs constitute the weakest section of India's population from the ecological, social, cultural, economic and educational point of view. Special intensive efforts have been made for the overall development of such areas by combining, to the maximum extent, possible resources from different sectors with those separately allocated for the welfare and development of tribal communities in India. The factors responsible for change in socio-economic and cultural conditions of the tribal people are discussed below:

7.2. CONSTITUTIONAL SAFEGUARDS FOR ST

The founding fathers of the Constitution of India desired to secure justice, social, economic and political for all citizens. They realized that the inequitable forces embedded in the socio-economic system and also the political organizations, had resulted in deprivation and disadvantages for the poor and the weaker sections of the society. They, therefore, considered it necessary to provide specific safeguards in the Constitution for the development of weaker section as given below:

7.2.1. Educational and Cultural Safeguards

- **Article 25:** Freedom of conscience and free profession, practice and propagation of religion.
- **Article 29:** Protection of Interests of Minorities (which include STs).
- **Article 46:** Promotion of educational and economic interests of Scheduled Castes (SC), ST and other weaker sections. The State shall promote, with special care, the educational and economic interests of the weaker sections of the people, and, in particular, of the SCs and the STs, and shall protect them from social injustice and all forms of exploitation.
- **Article 350:** Right to conserve distinct language, script or culture. Instruction in mother tongue.

7.2.2. Social Safeguards

- **Article 15:** Prohibits discrimination against any citizen on grounds of religion, race, caste, sex or place of birth. 1. The State shall not discriminate against any citizen on grounds only of religion, race, caste, sex, and place of birth or any of them. 2. No citizen shall, on grounds only of religion, race, caste, sex, place of birth or any of them, be subject to any disability, liability, restriction or condition with regard to: a) access to shops, public restaurants, hotels and places of public entertainment; or b) the use of wells, tanks, bathing ghats, roads and places of public resort maintained wholly or partly out of State funds or dedicated to the use of general public.
- **Article 17:** Abolition of untouchability.
- **Article 19:** Protection of certain rights regarding freedom of speech, etc. 1. All citizens shall have the right to – (a) freedom of speech and expression; (b) assemble peaceably and without arms; (c) form associations or unions; (d) move freely throughout the territory of India; (e) reside and settle in any part of the territory of India; (g) practice any profession, or to carry on any occupation, trade or business.
- **Article 23:** Prohibition of trafficking in human beings and forced labour. 1. Trafficking in human beings and begar and other similar forms of forced labour are prohibited and any contravention of this provision shall be an offence punishable in accordance with law. 2. Nothing in this article shall prevent the State from imposing compulsory service for public purposes, and in imposing such service the State shall not make any discrimination on grounds only of religion, race, caste or class or any of them.
- **Article 24:** Prohibition of employment of children in factories, etc. No child below the age of 14 years shall be employed to work in any factory or mine or engaged in any other hazardous employment.
- **Article 39:** A - Equal justice and free legal aid. The State shall secure that the operation of the legal system promotes justice, on a basis of equal opportunity, and shall, in particular, provide free legal aid, by suitable legislation or schemes or in any other way, to ensure that opportunities for securing justice are not denied to any citizen by reason of economic or other disabilities.
- **Article 338:** National Commission or special officers for SC and ST.
- **Article 339:** Control of the Union over the Administration of Scheduled Areas and the Welfare of STs.
- **Article 342:** Appointment of a Commission to investigate the conditions of ST.

7.2.3. Economic Safeguards

- **Article 16:** Provides for opportunities for all citizens in matters relating to employment or appointment or post in favour of SCs and STs – 1. There shall be equality of opportunity for all citizens in matters relating to employment or appointment to any office under the State. 2. No citizen shall, on grounds only of religion, race, caste, sex, descent, place of birth, residence or any of them, be ineligible for, or discriminated against in respect of, any employment or office under the State.

- **Article 38:** State to secure a social order for the promotion of welfare of the people. 1. The State shall strive to promote the welfare of the people by securing and protecting as effectively as it may a social order in which social, economic and political justice shall inform all the institutions of the national life. 2. The State shall, in particular, strive to minimize the inequalities in income, and endeavour to eliminate inequalities in status, facilities and opportunities, not only amongst individuals, but also amongst groups of people residing in different areas or engaged in different vocations.
- **Article 39:** Certain principles of policy to be followed by the State – The State shall, in particular, direct its policy towards securing – a) that the citizens, men and women equally, have the right to an adequate means of livelihood; b) that the ownership and control of the material resources of the community are so distributed as best to sub serve the common good; c) that the operation of the economic system does not result in the concentration of wealth and means of production to the common detriment; d) that there is equal pay for equal work for both men and women; e) that the health and strength of workers, men and women, and the tender age of children are not abused and that citizens are not forced by economic necessity to enter avocations unsuited to their age or strength; f) that children are given opportunities and facilities to develop in a healthy manner and in conditions of freedom and dignity, and that childhood and youth are protected against exploitation and against moral and material abandonment.
- **Article 335:** Claims of SCs and STs to services and posts – The claims of the members of the SCs and the STs shall be taken into consideration, consistently with the maintenance of efficiency of administration, in the making of appointments to services and posts in connection with the affairs of the Union or of a State: provided that nothing in this article shall prevent in making of any provision in favour of the members of the SCs and the STs for relaxation in qualifying marks in any examination or lowering the standards of evaluation, for reservation in matters of promotion to any class or classes of services or posts in connection with the affairs of the Union or of a State.
- **Article 244:** Clause (1) Provisions of Fifth Schedule shall apply to the administration and control of the Scheduled Areas and STs.
- **Article 275:** Special funds by the Union Government to State Government for promoting the welfare of ST and providing them with a better administration.

7.2.4. Political Safeguards

- **Article 164(1):** Provides for Tribal Affairs Ministers in Bihar, Madhya Pradesh and Odisha.
- **Article 243:** Reservation of seats in Panchayats.
- **Article 330:** Reservation of seats for STs in Lok Sabha.
- **Article 332:** Reservation of seats for SC and ST in the Legislative Assemblies of the States.
- **Article 334:** 10 years period for reservation.
- **Article 337:** Reservation of seats for STs in State Legislatures.
- **Article 371:** Special provisions in respect of Northeastern States and Sikkim.

7.3. OTHER FACTORS RESPONSIBLE FOR DEVELOPMENT OF TRIBALS

7.3.1. Educational Facilities

The Government has taken measures to provide educational facilities to the tribals. The special commitment of the National Policy on Education, 1986 (revised in 1992) to improve the educational status of STs continues to be the major strength in launching special interventions and incentives to improve the accessibility for the tribals who live in the far-flung remote areas and remain isolated.

Emphasis is being laid on vocational and technical training, which is opened in tribal areas. According to these measures, concessions, stipends, books, stationery and other equipments are provided. Residential schools have been set up for them in tribal areas. The National Literacy Mission (NLM) was launched in 1988 and it aims at imparting functional literacy to adult non-literates in the age group of 15-35 years in the country.

In the field of higher and technical education, special provisions such as reservation of seats, relaxation in minimum qualifying cut-off percentages, remedial coaching and scholarships were being extended by the Department of Secondary and Higher Education. Similar concessions are also given to ST students for improving their skills in the upcoming/modern trades, which have better employability.

Government initiative for 'Education for all' (Sarva Shiksha Abhiyan) is also helpful to increase the literacy percentage in the tribal villages. Mid-day Meal Scheme (MDMS) programme was started by the government and is still continuing to attain the goal of universalization of elementary education.

Status of Education among STs

Education forms an important component in the overall development of individuals, enabling them to greater awareness, better comprehension of their social, political and cultural environment, and also facilitating in the improvement of their socio-economic conditions. These hold true in the case of the STs in India.

Literacy Rate (LR) is defined as the percentage of literates among the population aged seven years and above. Although literacy levels of STs have improved, the gap in literacy levels, both for tribal men and women, has not declined significantly. In fact the gap has increased between 1971-1991, falling thereafter, to a little above the 1971 level. Table 10 illustrates a comparison between the LRs of the total population and the STs, from Census 1961 to Census 2011.

Table 10: Comparative LRs of STs and total population (per cent)

Category / Census Year	1961	1971	1981	1991	2001	2011
Total population	28.3	34.45	43.57	52.21	64.84	72.99
STs	8.53	11.30	16.35	29.60	47.10	58.96
Gap	19.77	18.15	19.88	22.61	18.28	14.03

Source: Statistical Profile of Schedule Tribes in India, (2013), MoTA, Statistics Division, Government of India.

For the ST Population in India, the LR increased from 8.53% in 1961 to 58.96% in 2011, while the corresponding increase of the total population was from 28.30% in 1961 to 72.99% in 2011.

7.3.2. Scholarships

The post-matric scholarships enable a considerable number of SC students to obtain post-matric and higher level of education resulting in their overall educational and economic development. The objective of the scheme is to provide financial assistance to the ST students studying at post matriculation or post-secondary stage to enable them to complete their education.

The Scheme provides for 100% central assistance to the State Governments' and Union Territories' administrations over and above the respective committed liability of the State/Union Territory. The committed liability of the Northeastern States has, however, been dispensed with.

The Central Government awards scholarships to deserving students for higher studies in foreign countries. Seventeen and half per cent of the merit scholarships are granted by the Centre, to the deserving students of lower-income groups. This is the major step for the financial assistance for seeking education for the tribals.

7.3.3. Credit Support

The governments of different States are encouraging the development of cottage industries by providing loans and subsidies through various schemes. Multipurpose co-operative societies such as Large Agricultural Marketing Societies (LAMPS) provide credit in cash and kind. Credit support through National Bank for Agriculture and Rural Development (NABARD), Regional Rural Banks, Land Development Banks, and Nationalized banks help to establish new entrepreneurship and development of agriculture.

National Scheduled Tribes Finance and Development Corporation (NSTFDC) is an apex organization set up for economic development of STs. This Corporation was incorporated as a government company under MoTA and granted license under Section 25 of the Companies Act, 1956. It is managed by the Board of Directors with representation from the Central Government, State channelizing agencies, Industrial Development Bank of India (IDBI), Tribal Co-operative Marketing Federation of India Ltd. (TRIFED) and eminent persons representing STs, etc. The Corporation helps in an economic upliftment of the STs by providing financial assistance at concessional rates of interest.

7.3.4. Financial Assistance

Financial assistance is given to the States and Union Territories for the construction of hostels and coaching of ST students for competitive examinations, upgradation of their merit, research and training, setting up of ashram schools, vocational training centres, free houses for tribals, village grain banks, educational complexes for ST girls in low literacy pockets and development of primitive tribal groups.

7.3.5. Marketing Support

Some tribal regions have no regular market. At the weekly *mandis* (market places), private wholesale dealers buy the minor forest produce at very low rates and sell them at huge profits in the cities. Many times, they pay the tribals in advance and later buy their goods at a nominal price. Since most of the tribals do not know how to count, the price that is promised often does not correspond to the amount they actually receive. Generally, the trader is a moneylender too, who buys the produce from the tribals as repayment of a debt, but at incredibly low prices. The Tribal Development Corporations, Forest Corporations and some other co-operative agencies have taken up the sale of some produce, but the moneylender-cum-trader is yet to be eliminated by them.

Prime objective of providing marketing assistance and remunerative prices to the tribals for their minor forest produce and surplus agricultural produce is to save them from exploitative middlemen and other functionaries. TRIFED was set up by the government in the year 1987. Tribal Research and Training Institute (TRTI) also started to give research and marketing support for the tribal communities.

The Ministry of tribal affairs provides grants-in-aid to its Corporation, TRIFED, to set off losses on account of fluctuations in prices of Minor Forest Produces (MFPs) being marketed by it for ensuring remunerative prices to tribals engaged in collection of MFPs either directly or through State Tribal Development Co-operative Corporations (STDCCs) and other such Co-operative Societies. NSTFDC provides term loan, bridge loan, marketing support loan, loan for Self Help Group, microcredit finance, educational loan, etc. for socio-economic development of the tribal people.

TRIFED continued to market tribal products through the network of its retail outlets 'TRIBES INDIA' in the country and during the year sold products worth Rs. 1,490.80 lakhs (Annual Report - 2013-14, MoTA, GOI).

Market development of tribal products/ produce and grants-in-aid to STDCCs for MFP operations (both central sector schemes) and schemes has been reviewed by merging the two schemes namely, Grant-in-Aid to STDCCS for MFP Operations, and Market Development of Tribal Products/ Produce under the scheme named as 'Institutional Support for Development and Marketing of Tribal Products/Produce'. The objective of the Scheme is to create institutions for the STs to support marketing and development of activities they depend on for their livelihood. These are sought to be achieved by specific measures such as:

- Market intervention
- Training and skill upgradation of tribal artisans, craftsmen, MFP gatherers, etc.
- Research and development/IPR activity
- Supply chain infrastructure development

7.3.6. Agriculture Development

Some of the major steps have been undertaken to increase production of agricultural crops such as providing irrigation facilities and land development, distribution of hybrid seeds, fertilizers and pesticides, etc. on subsidized rate, to provide agricultural implements and bullocks to the tribal people and also to provide high-yielding milch and other animals on subsidized rate. Like the cultivators of the advanced societies, some of the tribals in India are resorting to permanent-settled agriculture. The *Oraon*, *Munda*, *Gond*, *Bhumij*, *Ho*, *Santal* are efficient cultivators at present.

7.3.7. Allotment of Wasteland

Wasteland is allotted by the government for the cultivation of agriculture crops and tree crops for landless tribals on a permanent basis.

7.3.8. Social Land Allocation System

Social land allocation system helps to involve the local people in plantation programme. Rights on land may be given on a short-term basis, as in *Taungya*. People are allowed to raise agriculture crops in new plantation areas in return for their works in planting the tree and protecting them in the initial years. After this period, when the trees begin to shed out the agriculture crops, the farmers are moved on to a new piece of land.

7.3.9. Land Alienation

Land is the most important source of tribal livelihood for agriculture; both for settled and shifting cultivation, horticulture, forestry and animal husbandry in mixed farming systems. The total effect of land transfers has been devastating to the fragile tribal economy. The massive inward migration of non-tribals has also changed the ownership of land in the Scheduled Areas to the disadvantage of STs. Through this policy, it will be ensured that illegal transfer of tribal land holdings is eliminated and that the alienated land is restored to the tribal. Various States have enacted laws to prevent alienation of tribal land.

7.3.10. Tribal-Forest Interface

Various steps are implemented by the government to improve the natural resource base so that the socio-economic conditions of STs improve, including recognition of their age-old occupation rights and concession, ownership over forest produce, conversion of forest villages into revenue villages, collection of minor forest produce from forest areas, and rights for collection, handling and transportation of forest produce, etc.

7.3.11. Transport and Communication Facilities

In the successive development programs for the forest regions, top priority is accorded to the development of transport and communication facilities so that education, health, land colonization, housing, development of horticulture, animal husbandry and co-operative schemes can be initiated in the region, bringing the forest dwellers into the mainstream of nation development.

7.3.12. Health and Family Welfare

The National Health Policy, 1983 (being revised), categorically emphasizes the urgent need for improving the tribal health, especially through detection and treatment of endemic and other diseases specific to tribals.

State Governments introduced schemes for compulsory annual medical examination of ST population in rural areas. Under these schemes, Mobile Health Checkup Teams are deputed to villages according to a schedule drawn up annually. In case of a need for further investigation or treatment, tribal patients are entitled to avail of free facilities in government / referral hospitals. To enable the healthcare services to reach the STs, especially those living in the most backward remote areas, 52 districts in 13 States (Andhra Pradesh - 6, Bihar - 6, Gujarat - 3, Kerala - 5, Madhya Pradesh - 4, Maharashtra - 6, Manipur - 4, Odisha - 8, Rajasthan - 2, Tamil Nadu - 2, Tripura - 4, Uttar Pradesh -1 and West Bengal - 1) were identified by the Central Planning Committee. The State Governments were also asked to take special steps to check deaths of children due to malnutrition, epidemics, etc. in identified areas by the Central Planning Committee during summer/monsoon seasons by establishing adequate number of sub-centres, Primary Health Centers (PHCs) and Community Health Centers (CHCs) as per the relaxed norms in tribal areas. Besides, a focused attention was also paid to the deployment of medical and paramedical personnel in line with the recommended staffing pattern and regular field visits by them, stocking of essential medicines/drugs, provision of Mobile Health Units where feasible and chlorination of wells, etc.

Along with the National Malaria Eradication Programme, the other programmes to control Filaria, Japanese Encephalitis and Kala-azar were also implemented by States/Union Territories. The National Leprosy Eradication Programme was started in most of the States/Union Territories to address the problem of high incidence of Leprosy amongst tribals. Amongst the tribals, PTGs and the nomadic groups are passing through the most fragile health conditions, when compared to the others. Therefore, a new scheme called 'Medical Care for Remote and Marginalized and Nomadic Communities' was launched during the Ninth Five Year Plan. The programme of Reproductive and Child Health (RCH), which takes care of the maternal and child health needs, also made some special provisions for those living in remote areas, where the existing services at PHC level are under-utilized.

7.3.13. Employment

The forest department employs labourers for a variety of tasks such as nursery management and silviculture operations, which include plantation, weeding, fertilizer application, irrigation, soil working, training, pruning, plant protection measures, etc. Workers are also employed for departmental construction such as bridges, culverts and causeways, forest roads, forest office, guest house and road repair. Tribals are the prominent labour source in logging operation, which includes marking of tree, felling of tree, conversion, transportation, loading and unloading, etc. This gives much needed additional income to the tribals. With the expansion and the development of forest-based industries and cottage-based industries, the employment opportunities have been increased for the tribals in tribal areas.

7.3.14. Research

Research on ethnic profiles of the tribals, spectrum of problems and prospects, developmental constraints, and monitoring and evaluation of schemes and projects helps for effective implementation of different schemes and plans for socio-economic development. The existing tribal research institutes located in different States help to carrying out purposeful research and evaluation studies, and work towards the preservation of the rich tribal cultural heritage.

7.3.15. Reservation in Employment and Education

The most important stated aim of the Indian reservation system is to boost the opportunities for improved social and instructive position of the under-privileged communities, and thus, allow them to take their equitable place in the conventional Indian society. Impact of reservation policy on employment and education is positive. The strategy of reservation has had a helpful effect in conditions of induction of STs into public sector employment and in educational institutions.

7.3.16. Abolition of Bonded Labour

The bonded labour system (Abolition) Act is passed in year 1976. This Act is meant for the abolition of bonded labour system with a view to prevent the economic and physical exploitation of the weaker sections of the people and for matters connected therewith. Trafficking in human beings and beggars, and other similar forms of forced labour are prohibited and any contravention of this provision shall be an offence punishable in accordance with law. Article 23 of Constitution of India is on the prohibition of human trafficking and forced labour to protect the tribals from forced labour.

7.3.17. Village Grain Bank

This scheme provides grants for establishment of Village Grain Banks to prevent deaths of STs, especially children in remote and backward tribal villages facing or likely to face starvation, malnutrition and also to improve nutritional standards. The scheme provides funds for building storage facility, procurement of weights and measures and for the purchase of initial stock of one quintal of food grain of local variety for each family. A committee under the Chairmanship of the village headman runs the Grain Bank thus established.

7.3.18. Development of Handicrafts

Many subsidiary occupations including handicrafts are undertaken in the various tribal regions. These include basket making, spinning and weaving. The *Maria Gond* distils spirits from forest produce. Functional classes among the *Saora*, the *Kond* and the *Gond* devote themselves to cow-herding, metal working, weaving, can working, pottery and so on. The *Korwa* and the *Agaria* are well-known iron-smelters; producing tools for local use only, and their techniques being very crude. The *Ghasi* made gut from the fibrous tissues of animals. The *Tharu* depend upon farming, manufacture of furniture, household utensils, basket, musical instruments, weapon, rope and mats. The Madras *Irula* also makes bamboo mats and baskets, and ploughshares and wheels for local use only.

7.3.19. Women and Child Development

The National Nutrition Policy, 1993 recognizes the problem of malnutrition and undernutrition prevalent amongst tribal women and children, and strongly advocates the need for controlling the same. In pursuance of the commitments of the Policy, the Department of Women and Child Development through its nationwide programme of Integrated Child Development Services (ICDS) continued to provide the much needed nutritional and health inputs/services for the benefit of tribal children, adolescent girls and expectant and nursing mothers living in the remote tribal areas with relaxed norms. Of the total 4,608 ICDS projects in action by the end of the Ninth Plan, 758 (13.4%) were tribal projects through which a package of six services viz., health check-ups; immunization, supplementary feeding; referral services; non-formal pre-school education, and health and nutrition education were being extended to 4.77 million children and 0.96 million mothers. The concept of mini-*anganwadis* introduced in the tribal areas was only to ensure that ICDS services reach tne tribal women and children even in the remotest tribal areas.

7.3.20. Rural Development

As majority of tribals live in abject poverty, the Ministry of Rural Development plays a vital role in raising their status above the poverty line through implementation of various poverty alleviation programmes, and providing them with financial and other support for taking up self-employment and income-generation activities. In addition to the poverty-alleviation programmes, the Ministry also provides basic amenities such as housing, drinking and water.

The integrated programme of Swarnajayanti Gram Swarozgar Yojana (SGSY) and Jawahar Gram Samriddhi Yojana (JGSY) provides wage employment to STs/SCs. The two schemes i.e., JGSY and the Employment Assurance Scheme (EAS), were brought under the purview of the mega scheme of Sampoorna Grameen Rozgar Yojana (SGRY) in September 2001. The SGRY focuses on the generation of wage employment, creation of durable rural assets and infrastructure, and provision of food security to the rural poor.

Under the Indira Awas Yojana (IAY), about 60% of the total allocation was earmarked for STs and SCs. Similarly, Accelerated Rural Water Supply Programme, and Central Rural Sanitation Programme were also implemented in the tribal areas, especially for sanitation purpose. Under the National Old Age Pension Scheme, 2.4 million STs were covered during the Ninth Plan, accounting for 7.4% of the total beneficiaries. Similarly, while one lakh ST families were benefited under the National Family Benefit Scheme accounting for 10.2%, four lakh ST women were benefited under the National Maternity Benefit Scheme (NMBS) accounting for 7.4% of the total beneficiaries during the Ninth Plan (NMBS now stands transferred to the Department of Family Welfare with effect from 2001-02).

7.3.21. Minimum Support Price for Minor Forest Produce

A new scheme has been introduced by the Government of India for tribal and forest dwellers of the country i.e., 'Mechanism for Marketing of Minor Forest Produces (MFPs) through Minimum Support Price (MSP) and Development of Value Chain for MFP'. The scheme has been started with the objective of providing fair price to MFP gatherers, enhance their income level and ensure sustainable harvesting of MFPs. The MSP scheme seeks to establish a framework to ensure fair prices for the produce collected by them, assurance of buying at a particular price, primary processing, storage, transportation, etc. while ensuring sustainability of the resource base. It is a holistic scheme for development of MFP trade and would initially cover non-nationalized / non-monopolized MFPs from the 12 main identified MFPs namely, Tendu, Bamboo, Karanj seed, Mahua seed, Sal leaf, Sal seed, Lac, Chironjee, Wild Honey, Myrobalan, Tamarind, Gums (Gum Karaya). To start with, the scheme is being implemented in 102 districts of eight States namely, Andhra Pradesh, Maharashtra, Odisha, Chhatisgarh, Madhya Pradesh, Jharkhand, Rajasthan and Gujarat. It is also expected to support primary value addition of MFP, provide for supply chain infrastructure such as cold storage and warehouses, and emphasize scientific harvesting of MFP.

The scheme seeks to institutionalize various mediums and long-term aspects of sustainable collection, value addition, market infrastructure development, knowledge base expansion of MFPs, market intelligence development, etc. Strengthening the bargaining power of Gram Sabhas/ Panchayats and their involvement shall also be addressed and taken care of during implementation of the scheme. Government of India has extended financial support towards part of working capital requirement of State Implementing Agencies and also is sharing losses, if any, with the State Government in the ratio of 75:25.

Every year MSP announces for the selected MFPs with the technical support from TRIFED. TRIFED will acts as the Central Nodal Agency for implementation and monitoring of the scheme through State level implementing agencies. State designated agencies will undertake procurement of notified MFPs directly from MFP gatherers (individual or collectives) at haats – notified procurement centers – at grass root level at prefixed MSP, and ensure full and timely on the spot payment to MFP gatherers. The entire operation shall be monitored through State level committee headed by Chief Secretary and district level committees headed by District Collector for an effective implementation of the provisions of the scheme.

TRIFED has notified constitution of MFP Pricing Cell in September-2013. Based on the recommendation of the Pricing Cell, the MSPs for following 11 MFP items (Table 11) under the Scheme were announced by MoTA for implementation in the year 2013-14:

Table 11: MSP for Non-timber Forest Products (NTFPs)

Sr. No.	MFP	Price (Rs./ kg)
1.	Tamarind	22
2.	Honey	132
3.	Gum Karaya	108
4.	Karanj seed	21
5.	Sal seed	10
6.	Mahuwa seed	22
7.	Sal leaves	21
8.	Chironjee pods with seeds	100
9.	Myrobalan	11
10.	Lac - Rangini	210
11.	Lac - Kusumi	290

Source: Annual Report - 2013-14, MoTA, GOI

7.3.22. Tribal Sub-plan (TSP)

The present TSP strategy was initially developed by an Expert Committee set up by the Ministry of Education and Social Welfare in 1972 under the Chairmanship of Prof. S.C. Dube for the rapid socio-economic development of tribal people and was adopted for the first time in the Fifth Five Year Plan. The TSP strategy, with some modifications, continues till this day (Table 12) and the salient features with respect to TSP for States, are given below:

- The funds provided under the TSP of the State have to be at least equal in proportion to the ST population of each State or Union Territory.
- Tribals and tribal areas of a State or Union Territory are given benefits under the TSP, in addition to what percolates from the overall plan of a State / Union Territory.

The Sub-plan should:

- Identify the problems and needs of the tribal people, and critical gaps in their development
- Identity all available resources for TSP
- Prepare a broad policy framework for development
- Prepare a detailed departmentwise plan
- Define a suitable administrative strategy for its implementation
- Specify the mechanism for monitoring and evaluation.
- The TSP strategy has been in operation in 22 States and 2 Union Territories. However, since 2003-04, funds meant for Union Territories are being provided for in the budget of Ministry of Home Affairs and the Ministry is not concerned in the administration of funds in the Union Territories.
- TSP concept is not applicable to the tribal majority States of Arunachal Pradesh, Meghalaya, Mizoram and Nagaland, and in the Union Territories of Lakshadweep, and Dadra and Nagar Haveli where tribals represent more than 60% of the population, since the Annual Plan in these States/ Union Territories is itself a Tribal Plan.

Table 12: TSP States and Union Territories

Sr. No.	State/Union Territory	Sr. No.	State/Union Territory
1.	Andhra Pradesh	13.	Madhya Pradesh
2.	Assam	14.	Maharashtra
3.	Bihar	15.	Manipur
4.	Chhattisgarh	16.	Rajasthan
5.	Goa	17.	Sikkim
6.	Gujarat	18.	Tamil Nadu
7.	Himachal Pradesh	19.	Tripura
8.	Jammu and Kashmir	20.	Uttar Pradesh
9.	Jharkhand	21.	Uttarakhand
10.	Karnataka	22.	West Bengal
11.	Kerala	23.	Andaman and Nicobar Islands
12.	Odisha	24.	Daman and Diu

Source: Annual Report - 2013-14, MoTA, GOI

7.3.23. Scheduled Areas

The STs live in contiguous areas unlike other communities. It is, therefore, much simpler to have an area approach for development activities as well as regulatory provisions to protect their interests. Various provisions have been enshrined in the Fifth Schedule and the Sixth Schedule of the Constitution to protect the interests of STs with regard to land and other social issues. The Fifth Schedule under Article 244(1) of the Constitution defines "Scheduled Areas" as such areas as the President may by order declare to be Scheduled Areas after consultation with the Governor of that State. The Sixth Schedule under Article 244 (2) of the Constitution relates to those areas in the States of Assam, Meghalaya, Tripura and Mizoram, which are declared as 'tribal areas' and provides for District or Regional Autonomous Councils for such areas. These councils have wide ranging legislative, judicial and executive powers.

Purpose and Advantages of Scheduled Areas

- The Governor of a State, which has Scheduled Areas, is empowered to make regulations in respect to the following: 1. Prohibit or restrict transfer of land from tribals; 2. Regulate the business of moneylending to the members of STs. In making any such regulation, the Governor may repeal or amend any Act of Parliament or of the Legislature of the State, which is applicable to the area in question.
- The Governor may, by public notification, direct that any particular Act of Parliament or of the Legislature of the State shall not apply to a Scheduled Area or any part thereof in the State or shall apply to such area subject to such exceptions and modifications as he may specify;
- The Governor of a State having Scheduled Areas therein, shall annually, or whenever so required by the President of India, make a report to the President regarding the administration of the Scheduled Areas in that State and the executive power of the Union shall extend to the giving of directions to the State as to the administration of the said area.
- In accordance with the provisions of Para three of the Fifth Schedule to the Constitution, the States having Scheduled Areas viz., Andhra Pradesh, Chhattisgarh, Gujarat, Jharkhand, Himachal Pradesh, Madhya Pradesh, Maharashtra, Odisha and Rajasthan are required to submit the Governor's Report annually.
- The Tribes Advisory Councils (TACs) have been constituted in the nine Scheduled Areas States of Andhra Pradesh, Chhattisgarh, Gujarat, Jharkhand, Himachal Pradesh, Madhya Pradesh, Maharashtra, Odisha and Rajasthan. Though Tamil Nadu and West Bengal do not have any Scheduled Area, they have TAC. The direction of the Hon. President has been conveyed to the State of Uttarakhand in 2010 for the constitution of TAC in the State even though it does not have any Scheduled Area. Information about constitution of TAC in the State of Uttarakhand is awaited.

7.3.24. Cultural Tourism

Cultural Tourism (or culture tourism) is the subset of tourism concerned with a country or region's culture, especially its arts. It generally focuses on traditional communities who have diverse customs, unique forms of art and distinct social practices, which basically distinguishes it from other types/forms of culture. Cultural tourism includes tourism in tribal areas. It can also include tourism in rural areas showcasing the traditions of indigenous cultural communities (i.e., festivals, rituals), and their values and lifestyle. It is generally agreed that cultural tourists spend substantially more than standard tourists do. The reasons are not so hard to find because cultural tourists not only spend on food and lodging, but also on buying cultural artifacts and other cultural mementos, which are costlier in comparison to the mementos other tourists buy. This form of tourism can also be self-sustaining with a far more carrying capacity in comparison to other forms of tourism such as ecotourism and adventure tourism. This is because of the fact that cultural tourism is not seasonal in nature while the others are. Nowadays cultural tourism is recognized as a separate branch of tourism and special attention is being, and must be, given for the promotion of cultural tourism; if someone senses about sustainable tourism for economic as well as for tribal development.

7.3.25. Important Landmarks in Tribal Development

Recognizing the special needs and problems of tribals, a special niche was accorded to tribal development in the country's development agenda from the very beginning of the Plan Era. Some important landmark achievements in tribal development are as below:

- Programmes were designed with a special focus on STs (1951).
- Adoption of 'Panchsheel' – the five guiding principles of the process of tribal development (1956).
- Opening of multipurpose tribal development blocks for intensified development of STs (1961).
- Introduction of Special Strategies of TSP and Special Central Assistance (SCA) to TSP to ensure flow of population – proportionate funds from other developmental sectors for tribals (1974).
- Poverty alleviation programmes for at least 50% of the tribal families to cross poverty line and expansion of infrastructural facilities in tribal areas (1985).
- Setting up of special financial institutions viz., Tribal Co-operative Marketing Development Federation (1987) and National SCs and STs Finance and Development Corporation (1989).
- Ensure participatory development of STs at the grass root levels involving Panchayat Raj Institutions (PRIs) and Gram Sabhas as per the 73rd and 74th Amendment of the Constitution (1993) and the Panchayats (Extension to the Scheduled Areas) PESA Act (1996).
- A major shift in the approach from 'Welfare' to 'Development' and to 'Empowerment of Tribals' (1997); setting up of an exclusive MoTA (1999) and instituting a separate National STs Finance and Development Corporation (2001).

7.3.26. Government Policy, Plans, Schemes and Programmes for Development

The details of policy, plans, schemes and programmes implemented by the Government of India for the overall development of tribal community are separately discussed in Chapter 12 in this textbook.

8

FOREST AND TRIBES

8.1. INTRODUCTION

Tribal economy is connected intimately with forests. Forest, from the times immortal, has provided them with a deeply-loved home and these people have developed a well-balanced relationship with the environment in which they live. Dependency of the tribals on forest economy may be either partial or total. Forests are a vital livelihood support system for people in tribal area. Forest resources provide them with food, medicine, fodder and grass, fuel wood, and water for irrigation. Rights and concessions are given by the forest department to the tribal people for collection of forest produce from forest areas, and marketing support for their harvest.

8.2. ROLE OF FORESTRY IN SOCIO-ECONOMIC DEVELOPMENT OF TRIBES

8.2.1. Direct Employment

Forests provide a wide spectrum of livelihoods for tribal communities in the form of direct employment, self employment and secondary employment. The direct employment is provided by the forest department and other departments (rural development, agriculture and co-operatives) in the form of managerial, technical, research, planning and executive jobs.

The other direct employments consisted of labour force for rural masses generated by these departments under regular forestry activities such as growth, development and maintenance of the forests, research and training, survey of forest resources, protection and conservation of forest resources, soil and water conservation, civil construction such as forest road, bridges and culvert, harvesting, collection, processing and marketing of NTFPs, preparation of nurseries, fencing, soil working, transplanting, planting, tending operations, watering, fertilizer and pesticide application, protection and management of plantation and infrastructure development. This gives much needed additional income to the tribals.

8.2.2. Forest-Based Industries

The self employment in forestry creates livelihoods for local people through the sale of fuel-wood and fodder, grazing, lopping and grass cutting, forest-based handicrafts and cottage industries, sericulture, lac cultivation, bee keeping, charcoal burning, leaf plate making, liquor making, rope making and basketry, medicines, collection, processing and marketing of NTFPs, cultivation of agricultural crops under agri-silvicultural practices, livestock rearing, social and farm forestry, and availing of rights and concessions. The application of local skills and village-level technology in wood-based and small-scale forest-based enterprises provides secondary employment and livelihood opportunities for tribal people; main amongst them are saw milling, rayon, pulp and paper, ply wood and panel products, wood seasoning and preservation, tanning, sports goods, match splints, veneers, wooden boxes, bamboo and cane products, agricultural implements, furniture, structural wooden items, musical instruments, bidi making, educational goods, wood carving, wooden utensils, etc. Some of the tribes have started small-scale and cottage industries such as bamboo cane and allied products, hand-made paper, tendu leaves, distillation of essential oil, katha manufacturing, charcoal making, starch making, tassar product, bee keeping, lac propagation, broom making and miscellaneous products.

8.2.3. Collection of MFP

Forests provide significant social and economic benefits at all levels, especially in developing countries. Economics of people living in forest fringes has traditionally been dominated by subsistence agriculture.

NTFP such as fuel wood, medicinal plants, wild edible-vegetables, house building materials, fodder, fuel wood, mahua flowers *(Madhuca indica),* tendu leaves *(Diospyros melanoxylon)* wild mushrooms, tamarind *(Tamarindus indica)*, soap nuts *(Sapindus emarginatus)*, and wild brooms are an integral part of day-to-day livelihood activities, especially for tribal people. Forest products also can be used as raw materials for making a number of items for the home such as bidi, brooms, baskets, mats, rope, home-made toothbrushes and leaf plates. Some forest products such as charoli (*Buchanania lanzan*), lac (*Laccifer lacca*), dataoon as a teeth cleaning, leaf plates, and marula fruit (*Sclerocarya birrea)* are sold for a small cash income, which is sometimes used for buying clothing, oil or spices. These products may be used for subsistence or for sale, providing cash income.

More than 500 plants are being significantly used by the tribals as food, dyes, tannins, drugs, narcotic, drinks, housing instruments, weapons, fibres, medicine, etc. Thus, the NTFPs play a significant role in the livelihood of the forest dwellers and the communities living in the vicinity of the forest.

8.2.4. Forest Right Act

The STs and Other Traditional Forest Dwellers (Recognition of Forest Rights) Act, 2006, is a key piece of forest legislation passed in India on 18 December 2006. It has also been called the Forest Rights Act (FRA), the Tribal Rights Act, the Tribal Bill, and the Tribal Land Act. The law concerns the rights of forest-dwelling communities to land and other resources, denied to them over decades as a result of the continuance of colonial forest laws in India.

8.2.5. Ecotourism in Protected Areas

The country has a large treasure of natural beauty, and archaeological and architectural monuments. In addition, there are many hill and beach resorts, mountains, rivers, etc. The northern part of India is famous for its hills, plains, rivers, monuments, etc. – the Himalayas, Kashmir Valley, and the Ganges. There are colourful tribal lifestyles of the Northeastern States of Nagaland, Mizoram, Tripura and Manipur with their folk culture, which also is worth mentioning. In the Central Indian States of Odisha and Madhya Pradesh, tribal village life has resulted in a variety of artistically-executed handicrafts.

India has a huge ecotourism potential and this can be adjudged from this data that the country is having 634 nature-endowed areas, 103 national parks, and 531 wildlife sanctuaries (Ministry of Environment and Forest, 2015). Topography of homeland boasts of an abundant source of flora and fauna, and has numerous rare and endangered species in its surroundings. It is one of those rare countries, which has a wonderful blend of all the resources essential to make it as a golden star on the tourism map of the world.

Ecotourism, in simple terms, means management of tourism and conservation of nature in a way so as to maintain a fine balance between the requirements of tourism and ecology on one hand and needs of the local communities for jobs, new skills, income-generating employment and a better status for women on the other.

Tourism has long played an important role in the Indian economy as being the third most important industry sector of national economy. Within the tourism industry worldwide, ecotourism is one of the fastest growing sectors. The World Tourism Organization (WTO) has recently estimated that ecotourism is worth some $20 billion a year, and together with nature-based tourism, accounts for 20% of global international travel (WTO, 1998).

Positive impacts of tourism on indigenous populations in India in terms of economic growth and development (jobs and business opportunities); increased government expenditure on tourism infrastructure such as improvements of roads, water supply and sewage treatment, electrical facilities, education, communication that may also benefit residents and direct and indirect benefits due to tourism activities in tribal areas. Tribal people provide the services as guidance and interpretation, sightseeing, forest watcher, ecotourism enterprises such as sale of handicrafts, etc.

8.2.6. Joint Forest Management (JFM)

The most significant aspect of the National Forest Policy of 1988 is that it calls for the involvement of the forest-dwellers, tribals and the neighbouring communities of the forest in the management of the forests. Joint Forest Management (JFM) is a management tool intended to protect the forests and manage the non-wood resources of forests in a sustainable manner with the active participation of forest-dependent communities ensuring their livelihood needs. Forest-dependent families are the members participating in the planning and implementation of various forestry and community-development programmes.

JFM originated in West Bengal at the Arabari Forest Range in West Midnapore, near Midnapore town in 1971. The major hardwood of Arabari is sal; a commercially-profitable forest-crop was grown with active participation of the local people. A few years later in 1977, JFM was employed in the State of Haryana to prevent soil erosion and deforestation. The program led to reforestation of many hills in the State. After the initial successes in West Bengal and Haryana, the JFM schemes received national importance in the legislation of 1988 and thrust in the Guidelines of 1990.

As on March 2011, there were 1,18,213 JFM Committees (JFMCs) across 29 States and Union Territories of Andaman and Nicobar Islands, which are managing 22.94 million ha of forests in the country. Madhya Pradesh has the largest number of JFMCs among all the States with 1,228 JFMCs and also the forest land being managed under JFM with 6.69 million ha. A total of 14.5 million families are involved all over the country, which includes 4.6 million, ST families and 2.5 million SC families. About one-third of the land with the forest department is being managed today in partnership with the people of the country.

In India, JFM has emerged as an important intervention in management of forest resources. In many parts of India, small village groups have started to protect and reclaim degraded forest lands through collective action. The JFM Programme seeks to develop partnerships between local community institutions and State forest departments for sustainable management and joint benefit sharing of public forest lands. The primary objective of JFM is to ensure sustainable use of forests to meet local needs equitably while ensuring environmental sustainability. Distribution of area under JFM is presented in Fig. 28.

8.2.7. Development of Agriculture

Though the people had changed their cultivation pattern from traveller's cultivation to settled cultivation, agriculture is still the most primitive occupation of the tribal people. The technology is transferred to them without affecting their sentiment. A number of appropriate technologies are generated by the government and have become sustainable under their farming system.

8.2.8. Tribal Development Schemes and Programmes

Government of India implemented different schemes through forest departments such as National Afforestation Programme (NAP), National Bamboo Mission (NBM), schemes for the welfare of STs living within the forest, Tribal Insurance Scheme (TIS), Forest Right Act 2006, Green India Mission (GIM), gas cylinder on subsidy, ecotourism, National Medicinal Plant Board Scheme, schemes of social forestry such as *Kotwalia* schemes, *Malki* plantation, Food grain, and Social security scheme.

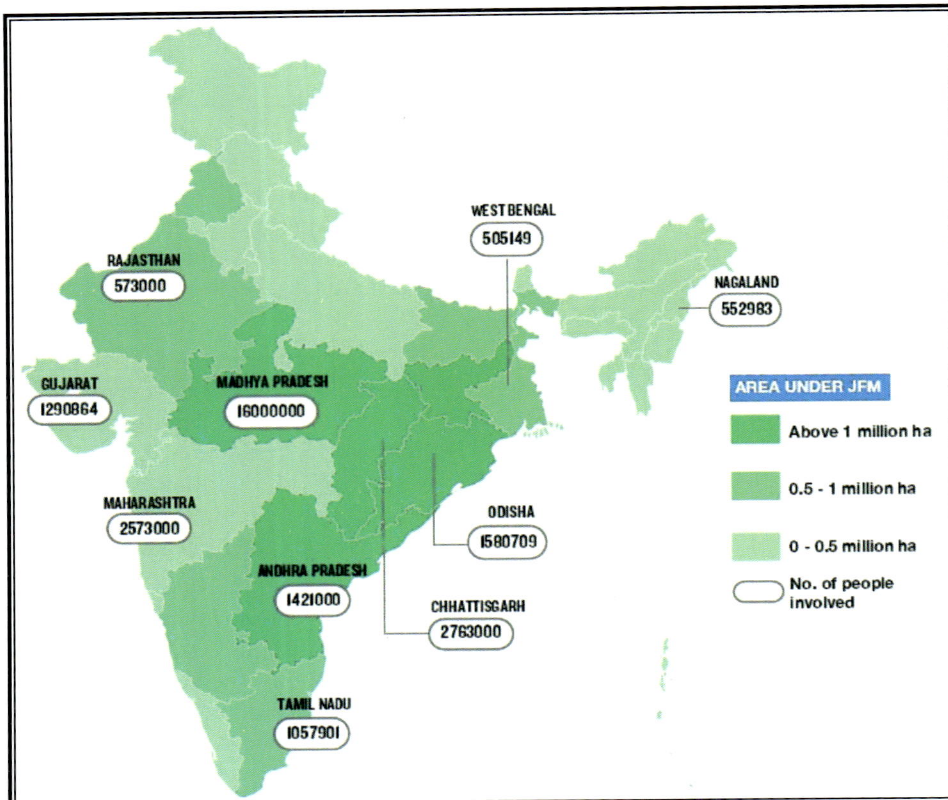

Fig. 28. Distribution of area under JFM in India (Source: ICFRE, 2011)

8.3. TRIBAL WELFARE AND SOCIAL FORESTRY

Social forestry means the management and protection of forests and afforestation on barren lands with the purpose of helping in the environmental, social and rural development. The term, 'social forestry', was first used in India in 1976 by The National Commission on Agriculture, Government of India. It was then that India embarked upon a social forestry project with the aim of taking the pressure off the currently-existing forests by planting trees on all unused and fallow land.

8.3.1. Objectives of Social Forestry

- To improve the environment for protecting agriculture from adverse climatic factors and thereby, reducing the pressure on the conventional forest area.
- To increase the supply of fuel wood, small timber for rural housing, fodder for livestock, and MFP for local industries.
- To increase the natural beauty of the landscape; create recreational forests for the benefit of rural and urban population.

- To provide jobs for unskilled workers and control the out migration.
- To check problems such as soil erosion, loss of biodiversity, depletion of ground water and over grazing.
- To raise the standard of living and quality of life of the rural and urban people.

8.3.2. Types of Social Forestry

Farm Forestry

Individual farmers are being encouraged to plant trees on their own farmland to meet the domestic needs of the family.

Community Forestry

Community forestry is a village-level forestry activity, decided collectively and implemented on communal land, where local populations participate in the planning, establishing, managing and harvesting of the forest crops. Hence, they receive a major proportion of the socio-economic and ecological benefits from the forest. Community forestry is a process of increasing the involvement of and reward for local people, of seeking balance between outside and community interests, and of increasing local responsibility. The animals are also allowed to graze in a particular plantation that is beyond damage, for maximizing the total output from the land.

Extension Forestry

Planting of trees on the sides of roads, canals and railways, along with planting on wastelands is known as 'extension' forestry. Extension forestry helps in increasing the boundaries of forests. Under this project, there has been creation of forests on the village common lands, government wastelands and *Panchayat* lands. Schemes for afforesting degraded forests of the government, which are close to villages, are being carried out all over the country.

Agro-forestry

Agro-forestry means practice of agriculture and forestry on the same piece of land. Agro-forestry is the growing of trees as an integral part of the cropping or livestock system. The home gardens of South-east Asia are one of the most highly-developed forms of traditional agro-forestry. The ecological and economical benefits derived from a combination of trees and crops are manifold.

8.3.3. Social Land Allocation Project

The land is allocated to the local people for plantation programme on a short-term basis. This helps to involve the people in plantation programme, reduce the cost of plantation and rehabilitation of the degraded lands.

8.3.4. Permanent Land Allocation System

The public land is allotted to individual cultivators on permanent basis for raising agricultural crops and forest crops.

8.4. SYMBIOTIC RELATIONSHIP BETWEEN FOREST AND TRIBES

Forests have played a vital role in the socio-economic and cultural life of the tribal people of India. It is well known that majority of the tribal population lived in isolation in the forested regions of India with harmony, security and trust for many centuries, and developed a symbiotic relationship with the forest.

The forest environment satisfies the deep-rooted tribal traditions and sentiments throughout their life until death. Even after death, the deceased tribal is laid to rest in a grave close to a forested region by most of the hunting and gathering tribes of India. Thus, even after death, the tribals are close to the forest.

Forests are the home of the tribals. They have been living in the forests ever since the evolution of mankind, and have adopted a balance with nature. Forest provides home, food and clothing. Plants and animals, and hence the human beings have intimate biological relationships since remote past and have evolved along parallel lines co-operating and depending upon each other for existence. Man and animals are dependent more upon the plants for their very basic needs viz., clean air, food, fibre, fuel and shelter.

Tribals have acquired practical knowledge about the ecosystems, functions, and the relationship between their societies and their living and non-living environment. The tribals derive very basic needs for survival from the biodiversity i.e., from themselves such as leaves, fruits, nuts, tuber, fodder for their livestock, fuel for cooking, ornaments, woods for fencing and tool making, mahua flowers, sal seeds, sal and tendu leaves, edible roots, tubers, bamboo and wild fruits, wild fauna, fish, fibre for clothing, rope, bark, bamboo and grasses for housing and farming, medicinal herbs for health care and diverse goods including gum, resin, dyes, wax, honey, and silkworm cacoons for their economy. In turn, they protect the forest, conserve it's fertility through their cultural activities, belief and practices.

Many rare and primitive cultivars of cereals, pseudo cereals, millets, pulses and vegetables, which have almost disappeared in modern society, are found still being grown by the primitive tribal communities living in accessible areas in the high hills of the Himalayas and Western Ghats of India. The primitive cultivars and wild relatives of the crop plants preserved by the tribals are some of the very rare and precious gene pool holding the genetic key to many valuable characters, which plants breeder may require for improving agronomical characters of the crop plants.

Thus, the tribals are exploiting the forest resources for their livelihood on one hand while on the other hand they are preserving, multiplying and conserving the valuable forest resources. Several tribal societies of India practice traditional farming, and still grow traditional crop plants and their wild relatives. They are most resistant to pest and fungal attack, and more adapted to withstand drought and waterlogging conditions, and have better value in terms of nutrient content.

Tribal belief and practices such as taboos and totem have helped to preserve flora and fauna in their pristine glory, representing a great ecological heritage for modern civilization. The tribals have acquired unique knowledge about the sustainable use of the wild flora and fauna by living close to the nature. Traditionally, the tribals do not commercialize the forest products i.e., wild plants and animal species, and never extract them in an unsustainable manner. It is to be noted that the tribal economy is basically subsistence economy. Traditionally, they produce or collect only for themselves and for their immediate community.

8.5. FOREST IN TRIBAL AREA

Tribal and their communities have been a part of forest ecosystem and their means and methods of livelihood have been deeply influenced by the forest. Forest plays a significant role in tribal economy and their socio-cultural practices are woven around forests. There are 189 tribal districts in 26 States/Union Territories as identified by the Government of India under the Integrated Tribal Development Programme (ITDP). Table 13 presents a summary of forest cover in tribal districts of the country as per India State of Forests (ISFR) - 2013 report published by Forest Survey of India, Dehradun.

Table 13: Forest cover in tribal district area in km^2

State	Number of hill/ Tribal district	Geogr-aphical Area (GA)	Very Dense Forest	Moderate Dense Forest	Open Forest	Total Forest	Percent age of GA	Change with respect to IFSR 2011	Scrub
Andhra Pradesh	8	87.090	239	16,465	8,359	25,063	28.78	-238	2.364
Arunachal Pradesh	13	83,743	20,828	31,414	15,079	67,321	80.39	-89	121
Assam	16	50,137	648	4,570	6,730	11,948	23.83	-48	93
Chhattisgarh	9	92,656	3,605	24,437	11,975	40,017	43.19	-40	87
Gujarat	8	48,409	322	2,937	3,512	6,771	13.99	5	395
Himachal Pradesh	3	26,764	950	1,067	1,218	3,235	12.09	4	118
Jharkhand	8	44,413	1,705	6,006	6,590	14,301	32.20	339	320
Karnataka	5	26,597	1,248	7,642	4,249	13,139	49.40	0	55
Kerala	9	27,228	1,147	6,846	5,414	13,407	49.24	311	29
Madhya Pradesh	18	1,39,448	5,631	20,235	16,362	42,228	30.28	-73	2,097
Maharashtra	12	1,44,233	7,261	11,775	11,665	30,701	21.29	-25	2.157
Manipur	9	22,327	728	6,094	10,168	16,990	76.10	-100	1
Meghalaya	7	22,429	449	9,689	7,150	17,288	77.08	13	372

State	Number of hill/ Tribal district	Geogr-aphical Area (GA)	Very Dense Forest	Moderate Dense Forest	Open Forest	Total Forest	Percent age of GA	Change with respect to IFSR 2011	Scrub
Mizoram	8	21,081	138	5,900	13,016	19,054	90.38	-63	0
Nagaland	8	16,579	1,298	4,736	7,010	13,044	78.68	-274	2
Odisha	12	86,124	5,249	14,356	14,237	33,842	39.29	544	2,472
Rajasthan	5	38,218	0	2,442	3,897	6,339	16.59	-10	903
Sikkim	4	7096	500	2,161	697	3,358	47.32	-1	311
Tamil Nadu	6	30,720	715	2,359	3,693	6,767	22.03	25	458
Tripura	4	10,486	109	4,641	3,116	7,866	75.01	-111	66
Uttar Pradesh	1	7,680	409	475	427	1,311	17.07	-8	1
West Bengal	11	69,403	2,957	3,709	7,880	14,546	20.96	2246	111
Aandman and Nicobar Islands	2	8,249	3.754	2.413	544	6,711	81.36	-13	57
Dadra and Nagar Haveli	1	491	0	114	99	213	43.38	2	1
Daman and Diu	1	72	0	1	3	4	5.01	-0.01	0
Lakshadweep	1	32	0	17	10	27	84.56	0	0
Grand Total	189	11,11,705	59,890	1,92,501	1,63,100	4,15,491	37.37	2,396	12,591

Source: India State of Forest Report (ISFR) - 2013, Forest Survey of India, Ministry of Environment and Forest.

9

CULTURE, SOCIETY, COMMUNITY, GROUP INSTITUTION, SOCIAL CHANGE AND CULTURAL CHANGE

9.1. MEANING OF CULTURE

Culture is one of the most important and basic concepts of sociology. In sociology, culture has a specific meaning. The anthropologists believe that the behaviour, which is meant, is called culture. In other words, the behaviour, which is transmitted to us by someone, is called culture. The way of living, eating, customs, language, religion, rituals, costumes, ornaments, singing, dancing and talking is all parts of a culture. Culture can also be defined as 'the behaviours and beliefs characteristic of a particular social, ethnic, or age group'. Culture has been defined in various ways by sociologists and anthropologists. Following are some of the important definitions of culture:

- **E.B. Tylor** defines 'Culture is that complex whole, which includes knowledge, belief, art, morals, law, customs, and any other capabilities and habits acquired by man as a member of society'.
- **Edward Sapir** says, 'Culture is any socially-inherited element of the life of man, material and spirituality'.
- **Malionwski** defines 'Culture, the handwork of man and conventional understanding manifest in art and artefact, which persisting through which he achieves his ends'.
- According to **Matthew Arnold**, a poet of the Victorian era, culture means 'contact with the best, which has been thought and said in the world'. He considered culture as a 'study of perfection'.
- **Pandit Jawaharlal Nehru** described culture as, 'the outcome and basis of training, establishment and development of physical and mental potentials'.
- **Sri. Rajgopalacharya**, the first Governor General of British India, defined it as, 'the collective expression of the thoughts, speeches and deeds of the learned, talented or creative members of a society or a nation'.

9.2. MEANING OF SOCIETY

The term, 'Society', is the most fundamental one in sociology. It is used in very wide sense. In common parlance, the society refers to members of specific groups, a number of persons associated for any temporary or permanent object; an association for mutual or joint usefulness, pleasure, or profit; a social union; a partnership. A society is a community or a group of people that are living together that often must live by the same set of rules and laws, and all have the same interest. Some definitions of the term 'society' are given below:

- **Maclver and Page** - 'Society is a system of usages and procedures, authority and mutual aid, of many groupings and divisions, of human behaviour and of liberties'.
- **Cooley** - 'Society is a complex of forms or processes each of which is living and growing by interaction with the other, the whole being so unified that what takes place in one part affects all the rest'.
- **Giddings** - 'Society is the union itself, the organization, the sum of formal relations in which associating individuals are bound together'.
- **Ginsberg** - 'A Society is a collection of individuals united by certain relations or modes of behaviour, which mark them off from others who do not enter into these relations or who differ from them in behaviour'.
- **Lapiere** - 'The term society refers not to group of people, but to the complex pattern of the norms of interaction, which arise among and between them'.

9.3. MEANING OF COMMUNITY

Community refers to a collection of people in a geographical area. It is also defined as a body of people having common rights, privileges or interests, or living in the same place under the same laws and regulations. Some definitions of the term 'community' introduced by sociologists are given below:

- **Aristotle**, who first defined the word 'community', as a group established by men having shared values.
- **Bogardes** defined it as 'the community is a social group with the same degree of we feeling and living in a given area'.
- According to **W. Ogburn**, 'community is a total organisztion of social life within a limited area'.
- **Davis** says, 'it is a smallest territorial group that can embrace all aspects of social life'.

9.4. MEANING OF GROUP

A collection of two or more individuals who have regular contact and frequent interaction, mutual influence, common feeling, and who work together to achieve a common set of goals.

9.5. MEANING OF INSTITUTIONS

Institution is an organization founded and united for a specific purpose. It is an establishment consisting of a building or complex of buildings where an organization is situated for the promotion of some cause. It is also understood that an organization, establishment, foundation, society, or the like, devoted to the promotion of a particular cause or program, especially one of a public, educational, or charitable character. For example, college is the best institution of its kind.

9.6. MEANING OF SOCIAL CHANGE

Social change refers to any significant alteration over time in behaviour patterns and cultural values and norms. By 'significant' alteration, sociologists mean changes yielding profound social consequences. Examples of significant social changes having long-term effects include the industrial revolution, the abolition of slavery, and the feminist movement. Social change is also defined as the change in social structure and relationship. It is a movement of people towards the establishment of environmental, economic and social justice, and the redistribution of wealth, power and resources.

9.7. MEANING OF CULTURAL CHANGE

Cultural change refers to a change in culture of the society through innovation, invention, discovery or contact with other societies. 'Culture change differs from other types of organizational change'. The shared beliefs, values, and behaviours of the organizational members become the target of the change process rather than solely focusing on the structures and systems within which people work.

9.8. DIFFERENCE BETWEEN SOCIAL CHANGE AND CULTURAL CHANGE

Table 14: Difference between social change and cultural change

Sr. No.	Social Change	Cultural Change
1.	Social change refers to the change in the social structure and relationship.	Cultural change refers to the change in the culture of the society.
2.	Change in culture brings about the change in a society.	Social change also brings a change in culture.
3.	Social values are values that a larger group holds - do not steal, do not murder, and do not cheat. These values can change as time goes on and society as a whole progresses (e.g., women's rights).	Cultural values are passed down from previous generations. Religion and how you treat the elderly are often cultural values. Cultures within society also have their own sets of norms. For example, drug culture, the gay and lesbian culture, university culture, etc. All these groups have their own set of norms within the society.

Sr. No.	Social Change	Cultural Change
4.	Social values are those that are mostly agreed by everyone regardless of the cultural values. For example, murder is regarded as wrong by most societies.	Cultural values are those that are held in a homogeneous population. For example, aboriginal peoples or closed-religious sects. These types of populations have a different set of values than those of the general population.
5.	Social change is a change in the way people communicate.	Cultural change is a change in all aspects of a person's or peoples' life, from currency, religion, government to the weather, fashion, art and sociology.
6.	Time and region - Social changes that persist over long durations of time and are imbibed, and generally accepted, lead to cultural change in society. Social change need not be restricted to any particular region. It can sweep across continents.	Cultural changes are more region-specific to the extent that people of a particular culture inhabit a particular geographic area.

9.9. CHARACTERISTICS OF SOCIAL CHANGE

9.9.1. Social Change is Universal

It means that social change is not confined to a particular society or group. It occurs in every society, sociologically speaking, an unchangeable society is considered as a dead society. Thus, no society is free from the impact of social change.

9.9.2. Speed of Social Change is related to Time Factor

The speed of social change is not uniform. It differs from period to period. In modern society, the speed of social change is rapid or faster than traditional society.

9.9.3. Speed of Social Change is unequal and Comparative

We can argue that speed of social change is more or less similar in each society. It is slow in traditional society whereas it is rapid in modern society. In urban areas, the speed of social change is faster than the rural areas.

9.9.4. Social Change is an Essential Law

Essential law means a law, which occurs and over which we have no control. It may be planned or not, but it must occur.

9.9.5. Social Change may be Planned or Unplanned

Planned changes are those, which occur by some deliberate or conscious effort. On the other hand, unplanned change refers to the change, which occurs without any deliberate effort such as earthquake, war, political revolution and other natural calamities. Thus, social change occurs in both, planned and unplanned manner.

9.9.6. Social Change may be Short Term or Long Term

Some change brings immediate change, which is known as short-term change, such as fashion and behaviour of the individual. Other changes take years to produce result, which is known as long-term change. Custom, tradition, folkways, etc. are long-term changes.

9.9.7. Social Change lacks Definite Prediction

Prediction means 'foretelling'. In case of social change, we are well aware of various factors, but we cannot predict although it is a law. Definite prediction of social change is not possible because we cannot say what will be the result of the social change.

9.9.8. Social Change is a Community Change

Social change does not refer to the change, which occurs in the life of an individual or life pattern of individuals. It is a change, which occurs in the entire community. This change can be called as social change, which influences a community form.

9.9.9. Social Change is the Result of the Interaction of Various Factors

A single factor can trigger a particular change, but never causes a social change. It is always associated with other factors such as cultural, biological, physical, technological and others. It is due to the material interdependence of social phenomena.

9.9.10. Social Change is Environmental

It must take place within a geographic or physical and cultural context. Both these contexts have impact on human behaviour, and in turn, man changes them. Social changes never take place in vacuum.

9.10. CHARACTERISTICS OF CULTURE

9.10.1. Culture is Learned

Learning a culture is the first essential characteristic of culture. Culture is learned by any individual, especially if he/she grows up in that particular culture. He/she can learn the behaviour patterns including language and many other attributes. We can learn the culture of others, especially if, we get acquainted with them and get influenced with their personality and lifestyle.

9.10.2. Culture is Shared by a Group of People

We can consider a thought or an action as a culture if, it is commonly shared and believed or practiced by a group of people. If it is commonly shared and believed by a group of individuals then it can be considered as their culture.

9.10.3. Culture is Cumulative

Knowledge embodied in culture can be passed on from one generation to another generation. More and more knowledge is added in the particular culture as time passes by. Every one may work out solutions to the problems in life that may be passed from one generation to another. This cycle remains as the particular culture grows with time.

9.10.4. Cultures Change

Not all the cultures accumulate there knowledge; thoughts or traditions are lost as new cultural traits are added. With this, there are possibilities of cultural changes with the particular culture through time.

9.10.5. Culture is Dynamic

It is considered that no culture remains in a permanent state. Culture is changing constantly as new ideas and new techniques are added in time, modifying or changing the old ways. This is the characteristic of a culture that stems from the culture's cumulative quality.

9.10.6. Culture is Ideational

Culture forms an ideal pattern of behaviours that are followed by individuals expectedly. It serves as guidelines that must be followed by the members to gain a social acceptance from the people with the same culture.

9.10.7. Culture is Diverse

It is considered that culture was formed by many separate cultures. It is a system that has several mutually-interdependent parts. Although these parts are separate, they are interdependent with one another forming culture as a whole.

9.10.8. Culture gives us a Range of Permissible Behaviour Patterns

Culture provides us with the norms of appropriate behavioural patterns. It involves how an activity should be conducted, how an individual should appropriately act as husband, wife, parents, child and others. It tells us how to dress based on gender, occasions, stature and many more.

10

TRIBAL RELIGION AND MAGIC

10.1. MEANING OF RELIGION

In every society, the existence of religion cannot be neglected. There is no society without religion. Even in this age of civilization, people are bounded by religion. Likewise in tribal society, the concept of religion was very much dominating. Thus, religion is the human response to the apprehension of something, or power, which is supernatural and supra-sensory. It is the expression of manner and type of adjustment affected by the people with their conception of the supernatural.

Religion is a collection of cultural systems, belief systems, and worldviews that establishes symbols that relate humanity to spirituality, and sometimes to moral values. Many religions have narratives, symbols, traditions and sacred histories that are intended to give meaning to life or to explain the origin of life or the universe. They tend to derive morality, ethics, religious laws or a preferred lifestyle from their ideas about the cosmos and human nature.

Religion is a fundamental set of beliefs and practices, generally agreed upon by a group of people. These set of beliefs concern the cause, nature, and purpose of the universe, and involve devotional and ritual observances. They also often contain a moral code governing the conduct of human affairs.

From the etymological point of view, Bouquet has shown that 'religion' is derived from the Latin word *religio,* which itself is derived from either root leg, which means to gather, count or observe, or from the root lig, which means to bind. In the former sense, the implication is belief in and observation of signs of divine communication. In the latter sense, the implication is the performance of necessary actions, which may bind together to the man and the supernatural power.

Following are some of the defination of religion given by different workers.

- Religion is the belief in spiritual being (**Edward Tylor**).
- Religion is a propitiation or conciliation of powers superior to man, which are believed to direct and control the course of nature and human life (**James Frazer**).
- Religion is a mode of action as well as a system of belief and sociological phenomenon as well as a personal experience (**Maliniwski**).

- Religion is the belief in one or the other super human or supernatural or supra-social power, which (the belief) has the fear, the reverence, devotion and the idea of sacredness for its basis, and which is expressed through prayer, worship or submission (**R. N. Mukharjee**).

Thus, from the above definitions, following conclusions are drawn:

- The concept of religion is prevalent in every society. In civilized society, the belief on supernatural power is still in existence.
- The belief regarding supernatural or supra-social power is found in religion and it varies from society to society.
- Rituals are performed in every religion. The exact reason is not known, but this may be due to the fear of supernatural power.

10.2. ORIGIN OF RELIGION

There are various theories about the origin of religion, following are main ones:

10.2.1. Animistic Theory of Edward Burnet Tylor

The scientific study of eighteenth century was diverted towards the analysis of primitive society and religion. Hence, many schools of thoughts were developed in the intellectual field. Out of these, animistic school is the main and N. S. Bergier was its pioneer. He believed that fetishism and astrology has grown out of a childish state of mind, which supports everything full of spirit. After that, various savage communities have become widely known to the civilized world. Economically and culturally, these communities were in horticultural stage by which it means that they were dependent chiefly on domesticated plants for their subsistence. Among them was found the very attitude of body, soul and anima.

Animism (from Latin *anima* 'soul, life') refers to the belief that non-human entities are spiritual beings, or at least embody some kind of life-principle. Animism encompasses the beliefs that there is no separation between the spiritual and physical (or material) world, and souls or spirits exist, not only in humans, but also in all other animals, plants, rocks, geographic features such as mountains or rivers, or other entities of the natural environment. Animism may further attribute souls to abstract concepts such as words, true names, or metaphors in mythology. Examples of Animism can be found in forms of Shinto, Hinduism, Buddhism, Pantheism, Paganism, and Neopaganism.

Animism is the belief that everything has a soul or spirit. An *anima* in Latin includes animals, plants, rocks, mountains, rivers and stars. Animists believe that each anima is a powerful spirit that can help or hurt them, and are to be worshipped or feared or attended to in some way. Animism is a primitive religion whose adherents have for thousands of years deified animals, stars and idols of any kind, and practiced spiritism, witchcraft, divination and astrology. They use magic, spells, enchantments, superstitions, amulets, talismans, charms, or anything that they believe will help to protect them from the evil spirits and placate the good spirits that are found everywhere in everything.

Edward Tayor very systematically analyzed all of these and developed a theory, which is known as Animistis theory. Its principal formulations are as follows:

- According to this theory, primitive man forms the idea of soul on the basis of two biological problems. The first is the phenomenon of sleepiness and death, the second is the dream and vision. In the first biological condition when a man is in a state of sleep, his life is isolated from his body for the time being. Sometimes, even dead persons appear in the dream. Thus, they developed a concept of continuance of soul after death.
- Secondly, primitive man observed all the things on the basis of his own existence. He received the body and the soul of all other things by analogy with his own. He, thus, contributed to the conclusions that there is no difference between man and other things.
- Hence, the primitive man reached to the concept of pure souls and pure spirits. The spirits may have their effects on individual bodies. On the basis of good and bad categories, good spirits are beneficial and bad sprits are harmful.
- According to this theory, the primitive man realized various parts of the nature animated by such spirits. Hence, we started the worship of nature e. g., worship of water, river, sun, sea, serpents, land, etc.
- From this stage of worship, there developed a higher polytheism in which there appeared the sky god, rain god, etc. The other channels of such development represent those gods, which are connected with the particular stage and functions of human life.
- Further, it is notable that the animistic theory assumes that there is an upward development of mankind. It is because the period when this theory was developed was contemporary with the development of the theory of evolution.

10.2.2. Fetishism of Compte Auguste

According to Compte, the development of human society had undergone through three stages namely, theological or fictive, the metaphysical or abstract, and the positive or scientific. In the theological stage, universe was explained to be the result of person being. In the metaphysical stage, universe was explained in terms of abstract force. In the third stage, universe was explained in terms of experiments and facts. Again, the first theological stage is subdivided into three namely, fetishism, polytheism and monotheisms. Thus, according to Compte, the stage of fetishism is the principal source of religion in which worship was started. In other words, fetishism is nothing, but the worship of natural objects.

10.2.3. Manism of Ghost Theory of Herbert Spencer

Manism (from Latin *manes*, 'departed spirit, ghost') was a theory of the origin of religion briefly advocated in the late nineteenth century by the popular British philosopher Herbert Spencer (1820-1903). According to Spencer, all people have a belief in serving the self of the dead man. There is a belief that self exist for a considerable period after death. The ghost propitiation is observed at the funeral as well as after it. But besides the people of such belief, there are people who believe in ghost for the subsequent interval. In the end, there is a class of advanced people who worship ancestors with the assumption that in propitiation, there is a permanent existence of ghost. Again, there is a class of people who make a division of distinguished ancestors and undistinguished ancestors. This division, according to Spencer, is more marked where ancestors were leaders of conquering races. The religion originates from the different forms of ancestor worship.

10.2.4. The Nature Myth Theory

This theory was founded by Friedrich Max Muller (06 December 1823-28 October 1900), more regularly known as Max Muller, a German philologist and orientalist, one of the founders of the Western academic field of Indian studies and the discipline of comparative religion. According to this theory, fear and wonder of natural objects is the main source of religion. There were many names for one object and one name for many objects. It produced a confusion of names, and it happened to be a combination of several gods into many. These natural objects were regarded as active, living and luminous. Hence, sky, Earth, moon, and Sun were brought with the relation of primitive supreme beings. All impressive natural objects, for example, the thunder, storm, bursts of rain, Sun, moon, etc. created wonder and fear amongst the primitive people. Therefore, they started their worship, which is the main source of religion.

10.2.5. Totemic Theory of Frazer

Sir James George Frazer (01 January 1854-07 May 1941), was a Scottish social anthropologist influential in the early stages of the modern studies of mythology and comparative religion. He is often considered as one of the founding fathers of modern anthropology.

Totemism has experienced a great influence on the religion. For the first time, it was brought with connection to religion by T.M.F. Leman. Later, the idea that totemism itself is not a religion, but a starting point of religion, came into existence. In other words, totemism is the initial stage of all the religions and is the source of culture, moral and social organization.

10.3. RELIGIOUS PRACTITIONERS / LEADERS

Many religious practitioners have aimed to band together interfaith dialogue, co-operation and religious peace building. In all systems of religion, individuals are required to perform, or at least assist performance of certain activities by means of which supernatural beings (God, spirit or ghost) are, in some fashion, propitiated or influenced. In many societies, the prayer of individuals and other similar activities form the bulk of religious act, but there are few number of societies in which such behaviour is not supplemented and guided by religious practitioners. There are always some persons in the society who, by virtue of special training, personality characteristics or both, are regarded as more skilled than others in influencing or making contact with supernatural being these are known as religious practitioners. In the tribal society, two main practitioners are found – Shaman and Priest.

10.3.1. Shaman

Shaman is the man or woman who serves a society as part-time religious practitioner. A shaman is a person regarded as having access to, and influence in, the world of benevolent and malevolent spirits, who typically enters a trance state during a ritual, and practices divination and healing. Shamans perform a variety of functions depending upon their respective cultures such as healing, leading a sacrifice, preserving the tradition by storytelling and songs, fortune telling, and acting as a psychopomp (literal meaning, 'guide of souls').

The functions of a shaman may include either guiding the souls of the dead to their proper abode, and/or curing (healing) of ailments. The ailments may be either purely physical afflictions such as disease, which may be cured by gifting, flattering, threatening, or wrestling the disease-spirit (sometimes trying all these, sequentially), and which may be completed by displaying a supposedly-extracted token of the disease-spirit, or else mental afflictions such as persistent terror, which may be likewise cured by similar methods. The shaman also tries to cure the patients by uttering specific vocabulary, language showing peculiar signs on his face and singing.

10.3.2. Priest

A religious leader who is part of an organized religion is considered to be a priest or priestess. Of course, different religions have different terms for these individuals – they may be known as rabbis, ministers, mullahs, imams, or something else. These individuals are the keepers of the sacred law and tradition. They are mostly found in large-scale societies. Priests are authorized by priesthood, or some other religious organization, to perform religious rituals designed to influence the supernatural world and to guide the believers in their religious practices. Priests personally do not have supernatural powers at their command.

Priests are initiated and ceremonially-inducted members of an established religious organization i.e., they are members of a priesthood. The priest is a full-time practitioner.

Table 15: Difference between shaman and priest

Sr. No.	Shaman	Priest
1.	Part time specialist in religious functions	Full time religious practitioner
2.	He receives his power either through direct experience with supernatural beings or visions.	Priest receives intensive trainings and not inspiration like shaman.
3.	Shamanism involves direct contact with supernatural power.	Priest is not directly related with supernatural power.

Sr. No.	Shaman	Priest
4.	The shaman usually functions in small relatively-private ceremonies.	Priest also works at private levels, but many times at a large level i.e., in a group or society.
5.	Shaman is believed to possess unusual power and he can cure diseases.	Priest possesses specific qualities and they acquire knowledge of performing rituals. They do not possess unusual power.
6.	Shamans work with and for the people; often the tribes or small groups.	Priests work for a hierarchy, most often an organization e. g., church, temple.
7.	A shaman works with spirits often of nature, plants or creation.	A priest works only with one spirit, their concept of one god.
8.	The shaman gains authority from the success of his visions and / or methods they use to treat, heal and take care of the people.	The priest gains authority from other priests and from the structure they work within.
9.	The shaman must constantly grow, nurture their relationship with spirit and may completely change the visions, rules and ceremonies they work with, considering the times and the place and the people.	Priest will often, though not always, get to a point where he is granted authority and from that point on change very little and there is no challenge to these rules; he must uphold them.

10.4. MEANING OF MAGIC

Magic plays a very important role in tribal society and is considered to be the aggregate of superstitions. It consists of a practical side in which various actions are performed by a magician. Thus, as mentioned by Frazer, magic is an art and never a science because the idea of science lacks in the mind of magician.

Magic can be defined as a rite and verbal formula projecting human desires into the world on a theory of human control to some practical end. Magic may be defined as a set of beliefs and practices designed to control the visible or invisible world for specific purposes.

10.5. FORMS OF MAGIC

There are four different forms/types of magic namely, imitative, contagious, sequential, and divinative.

10.5.1. Imitative Magic

This occultist seeks to bring about an event by symbolically causing it to happen. This is one of the most common practices of magic used for killing or injuring the enemy by making his figure in clay. Sometimes an earthen image is prepared and then it is injured. It is believed that as the image gets injured, the person concerned with will also be injured.

10.5.2. Contagious Magic

Contagious magic is based on the idea of association. It is believed that by performing acts on something that is related to or has been in contact with a particular object or being, one may gain control over the original. Contagious magic assumes that things, which were once related or in contact, continue to remain connected. It means, whatever is done with the first that also affects the other, e. g., a hairs, fingernails, teeth, cloth and faces are used by the magician. If these things are destroyed, it means the person to whom they belong will also be injured or killed. For instance, someone could use your fingernail clippings in a magical ritual that would cause you to love them or to fall ill and die.

10.5.3. Sequential Magic

Sequential magic includes most beliefs and actions dealing with supernatural cause and effect, which are neither imitative nor contagious in nature. It operates by assuming that when one event occurs after another, the first must have caused the second and will continue to do so. Most superstitions can be classified as sequential magic. For example, breaking a mirror brings bad luck.

10.5.4. Divination Magic

Unlike imitative, contagious, and sequential magic, the magic of divination is not concerned with causing or preventing the occurrence of various events. Its aim is to predict. Towards this end, a large variety of devices and procedures are used. Palmistry, for example, uses the lines, size, and shape of the hand.

Table 16: Difference between magic and religion

S.N.	Religion	Magic
1.	A belief in one or more spiritual being	Direct result without intervention of spiritual being
2.	Collective activity with presence of sacred centres	Individual activity often secret
3.	Gathering for one object e.g, worship and communion	Individual malicious objectives
4.	The pursuit of ends approved by the society	Malicious and therefore socially condemned
5.	Social function	Individual function
6.	Well being of society	Well being of individual
7.	Prayer may even fail	Magic must get success
8.	Tendency to sacrifice one's own self	Tendency to overpower the supernatural
9.	Comes after magic	Comes before religion
10.	God understands the utterance of the mantras (spell).	The power of mantras (spells) has no effect if, wrongly uttered or used.

10.6. WITCHCRAFT

Witchcraft, also called witchery or *spellcraft,* broadly means the practice of, and belief in, magical skills and abilities that are able to be exercised individually, by designated social groups, or by persons with the necessary esoteric secret knowledge. Witchcraft is a complex concept that varies culturally and societally, therefore it is difficult to define with precision. The cross-cultural assumptions about the meaning or significance of the term should be applied with caution.

Witchcraft can be roughly defined as the power of a person to harm or influence nature through occult means. Witchcraft, in historical, anthropological, religious, and mythological contexts, is the alleged use of supernatural or magical powers. A witch is a practitioner of witchcraft. Historically, it was widely believed in early modern Christian Europe that witches were in league with the Devil and used their powers to harm people and property. Particularly, since the mid-twentieth century, 'bad' and 'good' witchcrafts are sometimes distinguished, the latter often involving healing. The concept of witchcraft as harmful is normally treated as a cultural ideology, a means of explaining human misfortune by blaming it either on a supernatural entity or a known person in the community.

11

ANTHROPOLOGY: MEANING, NATURE AND SCOPE, BRANCHES, STUDY METHODS AND ROLE IN TRIBAL DEVELOPMENT

11.1. MEANING

Anthropology is unique among the social sciences in exploring humanity in all its aspects, past and present. It provides a broad understanding of the similarities and differences in the human experience through time and space. The 'science of humanity' – Anthropologists study human beings in aspects ranging from the biology and evolutionary history of *Homo sapiens* to the features of society and culture that decisively distinguish humans from other animal species.

Anthropology is a scientific study of origin, behaviour, and physical, social, economic and cultural development, environmental and social relations, racial relationship, biological characteristics, social customs and beliefs of human being. It has origins in the humanities, the natural sciences, and the social sciences.

The term 'anthropology' is from the Greek word *anthropos* means 'man', understood to mean mankind or humanity, and *logia* means 'discourse' or 'study', and was first used in 1501 by German philosopher Magnus Hundt. According to Tonipart, anthropology is the branch of natural history, and it also deals with man and the races of mankind.

On the basis of the above two definitions, the following conclusions can be drawn:

- Anthropology, in contrast, combines in one discipline, the approaches of both the biological and social science.
- It studies man as a member of the animal kingdom and on the other hand, studies the behaviour of the man in a society.
- It does not limit to a particular group or period of history.
- It studies pre-historic period and compares with the period of civilization.
- The comparative study of civilizations' centre, its attention on the difference and similarities in culture to be noted among many human groups that inhabit the Earth.

11.2. NATURE AND SCOPE OF ANTHROPOLOGY

The basic concerns of anthropology are 'What defines human life and origins?', 'How are social relations among humans organized?', 'Who are the ancestors of modern *Homo sapiens*?', 'What are humans' physical traits?', 'How do humans behave?', 'Why are there variations among different groups of humans?', 'How has the evolutionary past of *Homo sapiens* influenced its social organization and culture?' and so on.

Anthropology studies the emergence and development of man from the physical, cultural, and social point of view. Physical emergence, evolution and growth are studied by physical anthropology. The emergence and development of man, as an artifier and his social evaluation, are studied by pre-historical and cultural anthropology.

With the inclusion body of generalizations on application, anthropology may be conventionally divided into four broad sections namely,

1. Physical anthropology – Study of human evolution and growth
2. Pre-history and cultural anthropology – Study of the world of a man
3. Ethnology – Study of racial and cultural distribution of man
4. Applied anthropology – Application of findings of physical and cultural anthropologist i.e., in ordering race and industrial relations, in colonial administration, in development and welfare programme of undeveloped countries

The scope of anthropology is as given below:

- To understand how these primitive societies adopted the new ways of living and difficulties they are facing.
- In the eighteenth century, these primitive societies discovered by the explorers, travellers, business and missionaries also attracted the attention of the political philosophy of Europe as the latter thought that these societies were examples of the state of nature before any man-made government came into being.
- To collect information about the origins of human institutions. The evolution has not operated only in the case of the physical aspects of mankind, but also in their social aspect. It was, therefore necessary to study how the present institution such as family and religion gradually developed.
- Only one aspect of mankind is not sufficient for drawing conclusions and generalizations. Therefore, entire study is essential, which includes anthropology.
- Primitive society is vanishing because of fast industrialization and urbanization. The study of primitive society will serve as benchmark work, which will be useful for further research worke.

11.3. BRANCHES OF ANTHROPOLOGY

11.3.1. Ecological Anthropology

Ecological anthropology attempts to understand the relative influence of environment on human societies and how it is used by different societies. The ecological perspective is based on the assumption that constant interplay takes place between man and his environment, and these two are not two distinct spheres. However, the ecological perspectives in anthropology recognize that environment and culture are not separate spheres, but are involved in a dialectic interplay or reciprocal causality. According to this view, sometimes, culture plays a more active role and sometimes, environment has the upper hand.

11.3.2. Socio-cultural Anthropology

Cultural anthropology is the study of the culture and social structure of a community or society. Cultural anthropology emphasizes the understanding of the total configuration and interrelationships of cultural traits, complexes and social relationships in a particular geographic environment and historical context. Thus, it is concerned with the influence of geographic and historical as well as social and psychological factors in the analysis of the development of a culture, its present characteristics and the change it is undergoing. Social anthropology is concerned with the manner in which people associate and group themselves while cultural anthropology is concerned with the habits and customs of the people. The most elementary form of social grouping is the family. The two kinds of human relationships that lay the foundation for the family are birth and marriage. The principles of descent and alliance constituting kinship are studied under socio-cultural anthropology. Each society and culture has its own unique notion of marriage.

11.3.3. Economic Anthropology

The economic anthropologists concentrate on the economic transactions and processes covering production, consumption, distribution and exchange of products in the primitive and peasant societies. They focus on the modes of exchange including ceremonial exchanges. The concepts of reciprocity and redistribution including the nature of trade and market systems are also studied. The processes of economic growth and development in societies in the light of socio-cultural factors that influence and determine the economic activity in the society are also closely studied.

11.3.4. Political Anthropology

The focus of political anthropology is on the political process and functioning of authority, law, justice and sanctions in simple societies, political organization in egalitarian and stratified societies, focus of power and leadership, political processes emerging among nations and complex societies, political culture and the nation-building processes.

11.3.5. Physical Anthropology

It is concerned with man as a biological organism. Physical anthropology deals with the origin of man, the comparative study of racial and sub-racial groups, the measurement and classification of anatomical differences among men, problems of heredity, the influence of differences in the physical environment on the human organism and the interrelationship between biological and cultural differences among men. The anatomical and morphological changes that have occurred over the past millions of years to produce this species of mankind are studied in detail by the physical anthropologists. Physical anthropology is concerned about the extent to which biological factors influence the nature, behaviour and culture of humans and population at large. It can be divided into the following sub-branches:

- Primatology
- Ethnology
- Human biology
- Palaeo–Anthropology
- Human Genetics
- Nutritional Anthropology
- Medical Anthropology
- Physiological Anthropology - i) Forensic Anthropology, ii) Dental Anthropology, iii) Anthropometry.

11.3.6. Archaeological Anthropology

Archaeological anthropology traces the origin, growth and development of culture in the past. By past, it is meant the period before history when man had not acquired efficiencies of written language to record the story of his life. The archaeologists try to reconstruct the events of man's past dating back to millions of years ago. Archaeology tells us about the technology used in the past by analyzing the tools people have left behind. On this basis, it can shed light on the economic activities of the people. The engravings on the pottery, jewellery, etc. reveal the artistic capacities of the people. Certain aspects of religious beliefs can also be guessed by observing burial sites and the articles kept there. The archaeological anthropology makes an attempt to understand the geological processes, particularly the climatic phases that have left evidences on Earth's surface. The archaeological evidences are found in abundance mainly in river terraces. The main methods of archaeologists are excavation to discover artefacts to assign an approximate time period and to build the cultural history of man's past based on that.

11.3.7. Anthropology of Religion

The anthropologists have formulated many theories on the origin of religion among people. Some of the major theories are animism, animatism and primitive monotheism. The beliefs in natural forces and supernatural forces and beings are being investigated. The religious traditions including the rituals, ceremonies among primitive and peasant societies are being investigated. Practices under taboos and totemism are also being studied in great detail. The status and role of shamans, medicine men, and priests as important personalities in the magico-religious fields are also studied. Witchcraft and sorcery are examined as important aspects of primitive magic.

11.3.8. Linguistic Anthropology

Linguistic anthropology is that branch of anthropology, which deals with language. It is concerned with the languages of all people, past and present, as it is the chief vehicle through which man preserves and transmits his culture from generation to generation. Linguistic anthropology seeks to understand the processes of human communications, verbal and non-verbal, variation in language across time and space, the social uses of language, and the relationship between language and culture. It identifies many subtle elements of the world's languages, and documents their structure, function and history. Linguistic anthropologists study unwritten languages and also written languages.

11.4. METHODS OF ANTHROPOLOGICAL STUDY

Presently, there are four methods of study of anthropology.

11.4.1. Field Method

This is one of the important methods of studying anthropology. The anthropologist selects the specific field of tribal people and stays in that area for long period to study their way of life in scientific manner. While studying field method, there are some important elements as follows:

Time of study - The duration of the study may be very long. Earlier the travellers and the merchants used to draw certain conclusions regarding the life of tribals. However, even after staying for a long period, conclusion can not be drawn if an anthropologist does not have a scientific approach.

Participant Observation

The tribals are illiterate; therefore, for studying their life, questionnaire method is not much useful. The researcher has to stay with them and participate in their social functions. Thus, he becomes a member of their tribe. He observes their day-to-day life, different occasions and rituals. While studying, the researchers can become biased, but they should give the factual accounts of their life.

Language Study

When a researcher has participated and observed the details of their life, it is essential to know their language. If the researcher is not aware about the language then he has to depend on the observation. This will, sometimes, mislead him. He can ask the questions and obtain the reasoning of every act by knowing the language. Moreover, the tribals feel affinity towards a layperson who can understand their language. However, detailed knowledge of the language is very essential otherwise there is every possibility of drawing wrong conclusions.

Field of Study and Choice of Informants

In social anthropology, the study of man in total is essential and therefore, it is difficult to keep restrictions on research workers to undertake only particular field of study. Thus, the research workers have complete liberty for selection of any field in anthropology such as family, marriage system, divorce, religious rituals, language etc.

While collecting information under field method, the research worker should try to collect the information from the person having knowledge about that society, even though the principle is that every member of the society is a potential member. The research worker can establish rapport with the help of such persons with the tribals. Sometimes, it is necessary to collect information from the priest, shaman, etc.

11.4.2. Historical Method

Another important method widely used by the anthropologist is the historical method. This method helps to understand the knowledge of pre-historical society. Thus, it helps to understand the various aspects of the present society. This method has served the guideline and on this basis, certain conclusions can be drawn about the past life of tribals. In this method, the research workers have to take the help of coins and other written historical articles.

The prominent workers who have taken the help were Robart Croeber and Clark Wissler. According to their opinion, the diffusion about the culture of two societies takes place and thus, there is a wide scope to draw the conclusions. It is very difficult to evaluate the utility of this method in anthropology because tribal society was illiterate and there are little written resources and proofs, and therefore, difficult to draw conclusions. Redcliff Brown has stated the following limitations of this study:

- The conclusion drawn from the historical method cannot be de-certained whether they are correct.
- The second defect is that the research workers depend more on imagination rather than actual experience and hence, basically makes this method defective.
- The sequence of development can be obtained from this method, but besides this, there is no practical utility.
- Conclusion could not be drawn because of absence of the script and other tribal literature.

11.4.3. Comparative Method

This method is used in sociology and anthropology from ancient period. In the eighteenth century method, the entire culture of mankind and its prominent characteristics, behaviour among the members can be studied pointing out their similarities and dissimilarities, which held in making generalizations about the existing condition. This method is very useful in comparing religious economic life, political and social, particularly in tribals. Thus, by studying critically, the principles of human behaviour are established. While drawing conclusions, a researcher should be free from bias and also treat himself without involving in that particular aspect of tribals.

11.4.4. Functional Method

This method was used in social anthropology since 1920. The supporter of this method had criticized the historical method. According to their opinion, there is no new study that studies historic facts and history of the society. However, attention should be given towards the principle of anatomy rather than the development of human body. Above view inferred that the social anthropologist gave more stress on functional method rather than the historical method.

According to the supporter of this method, every part of the society is dependent and interdependent on each other. Every part of the society has its own legitimate functions. The stability and organization are existing in the society due to this system. In short, this method is equally important in studying tribal as well as progressive society in the world.

11.5. STEPS IN ANTHROPOLOGICAL STUDY

Generally, there are about seven steps. It should also be noted that these steps are not fixed ones. Some steps may not necessarily be followed in some research projects.

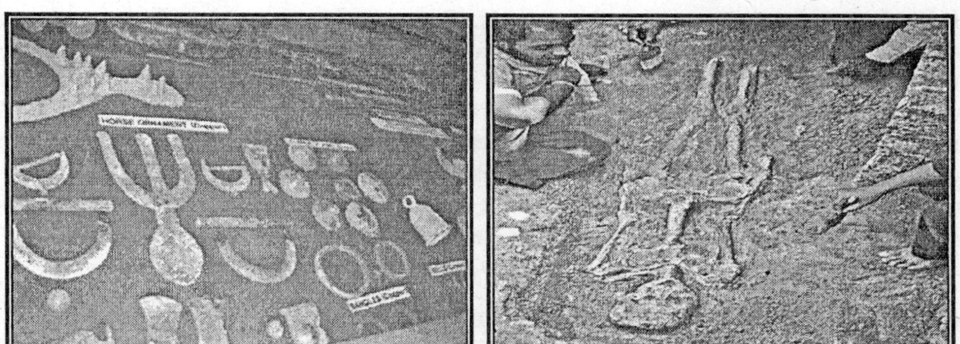

Fig. 29 Anthropological study

11.5.1. Identification of Research Problems

The first step is to come up with a research problem. Identification of research problem basically involves choosing a research topic. The ways and manners in which the researchers identify a research problem and choose a topic vary according to various factors. The research interests are often triggered by their own life experiences and observations. The initial ideas for research thus, may occur at any time and place for a researcher. Walking down a street, reading through newspapers, watching television, etc. may suggest a topic of research for an observing and curious person.

11.5.2. Literature Review

This step involves familiarizing or orienting oneself with the concepts, theories and the works already done pertaining to the topic identified. Available relevant literature on the topic chosen should be reviewed; we should also check out what works have already been done by others, what gaps are remaining, what questions remain unanswered, etc. Research work normally proceeds by reviewing earlier works on a specific research problem someone has identified. The researcher will need to review past works on the question he or she is raising. The traditionally-dominant source for literature review has been libraries and documentation centres where books and various references are found in card catalogued manner. Nowadays, most libraries maintain a computerized filing system, whereby references are made available via electronic online methods. Searching literature has become very easy, thus, with the computerization of library sources, one can easily access them, if internet connection is available. Literature review is necessitated by the fact that a researcher is probably not the first person to develop an interest in a particular problem; and hence, he or she needs to spend some time in the library reviewing the theories and methods others have used for this topic in the past and their findings.

11.5.3. Hypothesis Formulation

Hypothesis is a statement that can be proved to be correct or incorrect. Hypothesis formulation involves identifying basic research objectives and determining research questions. This should be tested empirically. We put some guiding assumptions to the research in this step. We ask some basic research questions. However, we may note that this may not be always the case. The type of research may determine whether hypothesizing is needed or not. For example, in exploratory studies, hypothesizing may not be needed.

11.5.4. Selections and Designing of Methods of Data Collection

Here, the researcher determines data collection methods and prepares data collection instruments. He/she chooses from among the different data collection methods. There are generally two categories of methods: Quantitative methods and qualitative methods. Quantitative methods focus on measuring quantity of information; terms such as prevalence, scope, percentage, frequency, magnitude, etc. are very important. On the other hand, qualitative methods focus on depth and quality of information. The complex, detailed and sensitive aspects, belief, attitudinal and knowledge dimensions, etc. are usually studied by qualitative methods.

11.5.5. Conducting Data-Gathering Activity

This is the step in which the researcher engages in collecting the needed data by using the various methods and instruments. The researcher goes to the field and collects the data. He/ she trains data collectors, supervises the overall data-collection process, and so on. Data collected, thus, may be of two types: primary and secondary data. Primary data are first-hand and original information; the researcher collects them first-hand. They are collected using research tools such as experiment, survey, questionnaire, interviews and observation. On the other hand, secondary data are those, which are already collected by someone else that are found in various sources as documents or archives.

They include official statistical documents, and mass media sources (such as electronic media – radio, television, films etc., and print media such as newspapers, magazines, journals, posters, brochures, leaflets, sign boards, etc.)

11.5.6. Data Organization, Analysis, Interpretation, and Report Writing

The most challenging task is how to manage, handle, store and arrange the raw data as cautiously as possible. Data may get lost, if not handled well. The researcher here, carefully stores the data, manages them, organizes and systematically arranges them. Various ways of analyzing data are used both in qualitative and quantitative methods. For quantitative data, researchers use sophisticated statistical techniques using computer models. Plans for data analysis are often made as early as before the data are collected. Analysis of qualitative data also actually begins while the researcher is in the field, recording his/ her field notes, tape recording and transcribing the interviews. Tape recording the interview process and transcribing are the essential components of analysis. In analyzing the data, the researcher must distinguish between his own views and the views of the people being studied.

11.5.7. Dissemination of Research Findings

This is the final step in which the researcher shares the findings with all the concerned bodies. Dissemination of the research findings is possible via scientific journals, seminars, symposiums, conferences and other forums. Steps may not necessarily be put in sequential order.

Fig. 30. Human development

11.6. ROLE OF ANTHROPOLOGY IN TRIBAL DEVELOPMENT

Anthropological studies play a fundamental role in the tribal development process. It is imperative to study the background, ethnic profile, cultural norms, behaviour patterns, attitudes, beliefs and practices before administering any intervention program or development plan. The development of tribal societies largely took place after World War II. After World War II, when former tribal societies were absorbed into new nations, 'applied anthropology' became 'development anthropology'. In the development era, it came to be assumed that economic development would automatically supply people with all goods and nourishment that they needed. Many applied anthropologists during the development era became involved in very small-scale remedial projects. However, anthropologists also study larger problems of cultural impact and social change associated with major development programs. Anthropologists have been trained and refined to have an eye to 'redefine an issue' i.e., collecting data to reveal particularly-problematic relationships, bureaucratic difficulties, or conditions harmful to tribal populations, to provide critical evaluation, and call public attention to the issues that have national or global impact.

Applied anthropologists work for groups that promote, manage and assess programs aimed at influencing tribal social conditions. Biological anthropologists work in public health, nutrition, genetic counselling, substance abuse, epidemiology, aging, mental illness, and forensics. Applied archaeologists locate, study and preserve prehistoric and historic sites threatened by development (Cultural Resource Management). Cultural anthropologists work with social workers, business people, advertising professionals, factory workers, medical professionals, school personnel, and economic development experts. Linguistic anthropologists frequently work with schools in districts with various languages. Anthropologists are the experts on human problems and social change; they can make a significant contribution in policy-affecting people. Professional anthropologists work for a wide variety of employers: tribal and ethnic associations, governments, non-governmental organizations, etc. Applying anthropological knowledge and data in development programs and bringing theoretical strategy into action is the prime focus of applied anthropologists, and is also the need of the hour for the progress of civilization, for justice, peace and harmony and for the oneness of tribal society.

Kroeber (1956) put forward the role of anthropology beautifully as compared to other disciplines in understanding humanity and implementing programmes for its betterment. If anthropology, an integrated science of man, deals 'holistically' both with the biological and cultural aspects of man, its applied or practical branch is equally comprehensive and strives to meet the need of mankind by planning for an all round development of human society. Commenting on the role of anthropology in identifying the problems and suggesting remedial measures for the human society, especially simple societies, George Foster puts forward three most important merits of anthropology:

1. Anthropology as a point of view
2. Anthropology as a storehouse or treasure of factual information
3. Anthropology's distinct research methodology

The holistic approach, belief in cultural relativism and a strong humanitarian tradition puts anthropology on a different pedestal. Anthropologists have accumulated a vast treasure of factual information by virtue of its vast experience of fieldwork for more than 150 years among the simple societies. The comparative method or cross-cultural comparison has added an important dimension to such studies. Among the tools of data collection utilized by the anthropologists, 'participant observation' is a unique way of getting the most reliable information described by many as 'insider's view'. Thus, anthropology, equipped with these distinct features and looking at 'development' through a different angle, may play a unique role in the tribal development.

12

POLICIES, PLANS AND PROGRAMMES OF TRIBAL DEVELOPMENT AND THEIR IMPLEMENTATIONS IN INDIA

12.1. PRE AND POST-INDEPENDENCE APPROACHES FOR TRIBAL DEVELOPMENT

In a democratic State, one group of people should not remain outside the mainstream society. Tribals have to be involved in all the matters concerning their country. After independence, planned development started, and broadly, three different approaches were used: 1. Isolationist, 2. Assimilationist, and 3. Integrationist.

12.1.1. Isolationist

The first approach was a legacy of the British regime, and is usually described as 'leave them untouched'. The policy was to isolate the tribal population from the masses. The British took deliberate efforts not to develop communication in the tribal areas. Tribals were kept away from the rest of the population. Verrier Elwin (1934) supported the establishment of a sort of 'National park' or 'specimens in a human zoo' of the tribals. He advised that their contact with the outside world should be reduced to the minimum, but this approach was not followed for long.

12.1.2. Assimilationist

The assimilationist approach is the approach, which paved the way for the tribal people to mingle with the neighbouring non-tribals. In India, the process of assimilation took place in different parts of the country, resulting in the gradual acceptance of Hindu culture by the tribals. The main criticizm against this approach was that this tried to change the tribals by imposing the non-tribal customs and traditions. The advocates of this view supported a direct assimilation without waiting for a slow and long-drawn change over. This approach is also considered to be a failure.

12.1.3. Integrationist

The past experience of the policies of isolation and assimilation forced the planners to take the middle way between the two, which is called the integrationist approach. This approach was mainly the brain child of Pandit Jawaharlal Nehru. The policy of integration consists of two types of measures for tribal development: (1) Protective measures, (2) Promotional measures.

The former consists of land policies, forest policies and the policies to protect the tribal culture and traditions. The latter is the same as development and welfare programmes undertaken by the government (through plans) and other voluntary agencies to make the tribal life better through the integrationist approach. P.D. Kulkarni states that 'the policy of protection and development is undoubtedly same in itself, but it remains to be seen whether development is possible without upsetting the harmony that exists in the placid tribal life.

12.2. CONSTITUTIONAL PROVISIONS

The Constitution of India initiated the tribal welfare programmes by incorporating important provisions. These provisions not only put an end to the policy of indifference, which had been followed during the pre-independence period, but also gave testimony to the tribal welfare programmes, making them a constitutional obligation for the future governments of the country. Major provisions are given as follows:

- Statutory recognition of tribal communities
- Creation of Scheduled Areas for the thorough development of the tribals
- Special representations in the Parliament, in the legislative assemblies and local bodies
- Special privileges in the form of reservation of a certain percentage of posts in government services and seats in educational institutions
- Recognition of the right to use local language for administration and other purposes, and to profess one's faith

In addition to the above, three provisions of the Constitution deserve special mention. According to the fifth schedule, Union Executive is given the power of giving direction to the States in matters relating to the administration of Scheduled Areas. The sixth schedule designates tribal areas in Assam and Meghalaya, where autonomous district councils and regional councils have been constituted, with powers to make laws for management of land, forests, shifting of cultivation, appointment or succession of chief and headman, inheritance of property, marriage and divorce, social customs and matters relating to village or town administration. Article 275(1) of the Constitution provides for grant-in-aid from the Union to the States for promoting the welfare of the STs or for raising the level of administration of the Scheduled Areas. The Constitution also provides for the appointment of a commission for STs for safeguarding their interests. Thus, tribal development, in the true sense and in the present context of India, is, in fact, a post-independence concept and draws the spirit from the Constitution itself.

12.3. SIVARAMAN COMMITTEE, U.N. DHEBAR COMMITTEE AND S.C. DUBE COMMITTEE

The National Committee on the development of backward areas under the chairmanship of Sri. Sivaraman (Sivaraman Committee) has recommended the 'Sub-plan approach' with suitable adaptation for other backward areas for better planning and development. Some committees were formed by the Government of India under the chairmanship of prominent politicians/government officials. These committees were included in the working group report of Fourth Plan 1980-85. The Scheduled Areas and SCs commission set up under the chairmanship of Sri. U.N. Dhebar in 1961 referred to this policy approval and observed that 'the problem is not to disturb the harmony of tribal life, but simultaneously work for their advance, not to impose any thing upon the tribals, but work for their integration as members and part of the Indian family'.

A study team appointed by the Planning Commission headed by P. Shilu A.O. opined that 'the aim of tribal welfare policy should be defined as the progressive advancement, social and economic betterment of tribals with a view to their integration with the rest of the community on a footing of equality within a reasonable distance of time'. The period has necessarily to vary from tribe to tribe and it may be five or ten years in the case of certain tribes.

The Dhebar Commission (1961) stressed an integrated approach. According to them, 'problem of economic development for the bulk of the tribals cannot be solved unless that resource of land, forests, cattle wealth, cottage and village industries are all mobilized in an integrated basis. In 1972, under the chairmanship of S. C. Dube, an Expert Committee was set up for advising on the formulation of a new strategy during the Fifth Five-Year Plan.

12.4. FIVE-YEAR PLAN

Planning is generally accepted in many developing countries as an indispensable means to promote development. The preference for it arose out of the inability of the poor and traditional societies to imitate and promote the development process on the one hand, and the desire of the State to put an immediate end to human sufferings on the other. The preparation and implementation of the plans of development in these countries respect the acceptance of the responsibility of development on the part of the States. During the process of planning for development in these countries, some new problems have emerged and the existing ones have assumed new dimensions leading to significant changes in the concept and strategies of planning and development. In India, planning processes are being implemented for over six decades. Tribes encounter socio-economic, cultural and political problems. They are considered as a weaker section of the society. The tribal development planning is being implemented along with Five-Year Plans under the control of the Government of India.

In the Five-Year Plans, the programmes for the welfare of the STs aim at:

- Raising the productivity levels in agriculture, animal husbandry, forestry, cottage and small-scale industries to improve the economic conditions

- Rehabilitation of bonded labour
- Education and training programmes
- Special development programmes for women and children

The Planning Commission has laid down the objectives and strategies for tribal development from time to time. The following are the account of the tribal welfare programmes in India during the various plans:

12.4.1. First Five-Year Plan (1951-1956)

The First Five-Year Plan outlined a positive policy for assisting the tribals as under:

- Assisting them to develop their natural resources and to evoke a productive economic life wherein they will enjoy the fruits of their own labour, and will not be exploited by more-organized economic forces from outside
- It is not desirable to bring about changes in their religious and social life, except at the initiative of the tribal people themselves and with their willing consent
- It is accepted that there are many features in tribal life, which should not only be retained, but also developed
- The qualities of their dialects, and the rich content of their arts and crafts also needs to be appreciated and preserved

Taking into consideration the conditions of the tribal people, the First Planning Commission quoted that 'There may be a good deal of justification for such (isolation) a policy of non-interference; but it is not easily practicable when tribal life has been influenced by social forces from without, and tribal communities have reached a certain degree of acculturization accompanied by the penetration of communications in the tribal areas, and of social services for the betterment of their lives'. In the First Five-Year Plan, Community Development Projects were started for all round development of rural areas, especially the weaker sections.

12.4.2. Second Five-Year Plan (1956-1961)

During this plan, the Ministry of Home Affairs provided fund to the Ministry of Community Development to establish Special Multi-Purpose Tribal Blocks (SMPT) in areas with prominent tribal population.

12.4.3. Third Five-Year Plan (1961-1966)

Towards the end of the Second Plan i.e., in 1959, the Government of India appointed a committee under the chairmanship of Verrier Elwin to review the SMPT Blocks. According to the recommendations of this committee, during the Third-Plan period, SMPT Blocks were renamed as Tribal Development Blocks (TDB) and suggested it to be opened in all the areas where over 60% of the population were tribals. In addition to the normal allotment of Rs. 12 lakhs to a community development block, a provision of Rs. 10 lakhs for first stage and Rs. 5 lakhs for second stage for TDB was also made.

12.4.4. Three Annual Plans (1966-1969)

During this period, no special funds were provided for tribal development. However, in 1969-70, a decision was taken to extend the total life of TDBs to 15 years by incorporating a new stage - three. During the third stage, each TDB was given Rs. 10 lakhs.

12.4.5. Fourth Five-Year Plan (1969-1974)

During the Fourth Five-Year Plan, a series of programmes were conceived and addressed to specific target groups. The Small Farmers Development Agencies (SFDA) and Marginal Farmers and Agricultural Labourers (MFAL) development agencies were the first two in the series. In these cases, attention was shifted from area development to development of identified individuals who qualified for special attention according to certain objective criteria. The Drought-prone Area Programme (DPAP) was another measure in the same direction, but with a difference. Here, the attention is given to the problem faced by an entire region, which is depressed because of its agro climatic situation. The specific target-group approach, however, was adopted to give attention to the weaker sections of the society. In the wake of establishment of these area-specific and weaker-group oriented projects, the programme for tribal areas were also on a pilot basis. Tribal Development Agencies (TDAs) were established on the pattern of SFDA, which addressed themselves to the problems of the tribal population. The level of investment in the new programme was much higher compared to TDB.

Six TDAs were started during the Fourth Plan. Each TDA covered a group of TDBs. TDAs were expected to comprise elements of economic development, social services and prospective measures. By the time of Fourth Plan, one of the drawbacks of the functioning of TDBs became clear that their activities were not properly integrated with the general development plans for the region. The Fourth Plan tried to rectify this drawback by adopting the integrated area development approach.

12.4.6. Fifth Five-Year Plan (1974-79)

During the middle of the Fourth Five-Year Plan i.e., in the year 1972, the Planning Commission set up a 'Task force on Development of Tribal Areas' with L.P. Vidyarthi as the Chairman. In their appraisal, the task force observed that in spite of various kinds of investments by the State and Central governments for the tribal development in successive plans, the problem of the tribals reflected in primitive methods of agriculture, land alienation, indebtedness, adverse effects of industrialization, low rate of literacy, poor health of nutrition, etc., had not been solved.

The committee opined that one of the important factors for the lack of impact so far, was the development of STs and tribal areas had been looked upon as a problem of 'welfare' as distinguished from 'development'. Taking into account the recommendations of the task force and other previous committees, during the Fifth

Five-Year Plan, an altogether new approach was adopted towards tribal development. This was termed as 'Tribal Sub-plan'. It envisaged the total development of the tribal areas and provided the mechanism for integrating the developmental activities of the government and the semi-government organizations by financing through the ITDP. The Sub-plan aimed at narrowing the gap between the levels of development of tribal and other areas, and to improve the quality of life of the tribal communities in general.

12.4.7. Sixth Five-Year Plan (1980-85)

The Sixth Plan continued the Sub-plan approach of the Fifth Plan. This was to be supplemented by target beneficiary approach with the objective of narrowing the gap between the level of development of the tribals and other developed communities, and bringing about a qualitative change in the life of a tribal community. The broad objectives of the Sixth Plan were:

- A progressive reduction in the incidents of poverty and unemployment
- Improving the quality of life through minimum needs programme
- A reduction in inequalities of income and wealth
- Infrastructure development for further exploitation of potential of the tribal region

The strategy of development laid emphasis on consolidation of the gains of protective measures, programmes of full employment, education and health services. The programmes under different sectors of development required to be intensified with suitable modifications to remove the present inadequacies in implementation. The States were to give due importance to the integration of programmes in the field and effective delegation of powers to the Project Authorities in ITDPs.

The approach in the Sixth Plan for the development of backward areas in general was to rely, to a greater extent, on the development of agriculture, village and small-scale industries, subsidiary occupations and related services, and also the Minimum Need Programmes and Area Development Programmes. Improvement of economic status of the tribals was the first concern, and suitable programmes of horticulture, cattle development, poultry and piggery were carried out. Emphasis was placed more on family-oriented programmes than on infrastructure development unlike in the previous plans.

12.4.8. Seventh Five-Year Plan (1985-90)

The basic premises of the TSP continued in the Seventh Plan also. During the Seventh Plan, the TSP strategy comprised the following:

- Identification of the development blocks, where the tribal population is in majority, and their constitution into ITDPs with a view to adopt an integrated and project-based approach for development.
- Marking of funds for the TSP and ensuring the flow of funds from the control of State plan, sectoral outlays and from financial institutions.

The programme of tribal development with ITDP pattern was continued in the Seventh Plan also without any basic or major changes in the approach, pattern or structure, but better co-ordination between various agencies, and social services was given priority. LAMPS were to be strengthened through broadening their popular base. Seventh Plan paid attention towards the rehabilitation of poor tribals and the removal of tribal women's backwardness.

12.4.9. Eighth Five-Year Plan (1992-97)

In the Eighth Plan, taking in view the problem related to the implementation of the schemes for tribal development, the Planning Commission tried to be more realistic. The plan largely emphasized the re-orientation of administrative structural levels for functional co-ordination, interrelation and effective delivery of services. The strategy of the Eighth Plan also specifically aimed at improving the living environment of the tribals by giving them better social and civic amenities and facilities. The working group has recommended that the objective of the Seventh Plan would continue for the Eighth Plan period. The objectives for the Eighth Plan are below:

- Progressive reduction in poverty and creation of employment thereby, providing reduction in income inequalities
- Improving the quality of life through a minimum needs programme
- Development and strengthening of infrastructure for further economic exploitation of the TSP area
- Development of confidence in tribals along the desired lines through intensive educational efforts

In the Eighth Five-Year Plan, TSP area, Modified Area Development Approach (MADA), Scattered Development Plans, and Primitive Tribe Development Plans for the tribal development approach have been stressed. Despite the efforts to diversify economic activities in non-farm sectors, the predominant source of livelihood in TSP areas continues to be agriculture. The main thrust would be on the development of fisheries, sericulture, horticulture, plantation on wasteland and growing vegetables. During the Eighth Plan, these areas would provide supplemental income and new avenues of employment to the tribals. Human resources development through education, vocational/craftsman training would be taken up to improve the skill of the tribals. Expansion of irrigation facilities and electrification of tribal settlements, expansion of irrigation wells, fertilizers, improvement of cattle breed and mining activities have also been given. In this plan, 66 family-oriented schemes have also been stressed to uplift the tribal families. The community development programmes are given second priority.

12.4.10. Ninth Five-Year Plan (1997-2002)

Ninth Five-Year Plan (1997-2002) has the strategy to achieve a seven percent growth rate for the economy. The main objectives of the Ninth Plan are the generation of adequate productive employment, eradication of poverty, empowerment of women, and socially-disadvantaged groups. It aims to ensure food and nutritional security for all, particularly the vulnerable sections of the society. The Ninth Plan, therefore, contemplates to implement people-oriented planning in which the government and the people, particularly the poor, can fully participate.

A participatory planning process is an essential precondition for ensuring equity as well as accelerating the rate of growth of the economy. For approaching to this goal, the Government of India enacted an amendment to the Constitution of India (73rd Amendment) for ensuring representation of backward and marginalized communities e. g., SC, ST and women in Panchayati Raj institutions. The draft report of the working group on decentralized planning and Panchayati Raj for formulation of the Ninth Five-Year Plan (1997-2002) noted that large parts of the country and vast section of the population have been untouched by this progress. It was now felt that the fruits of development could have been multiplied manifold with the involvement of people at the grassroot level in the process of development.

The approach for the Ninth Five-Year Plan has drawn attention to the need for national policy for tribal development. Efforts were taken to ensure that the tribal economy is protected and supported against threats from the external markets. The ownerships/patent rights of the tribal people in respect of MFPs vis-a-vis the use of medicinal plants were protected as per the provision of IPRs. Ninth Plan is a process and it would be able to fulfil the ambition of tribals is the initiation of twenty first century and may be able to bring them more nearer to the national mainstream.

12.4.11. Tenth Five-Year Plan (2002-07)

The Tenth Five-Year Plan lays down its first priority in finding solutions to these very unresolved issues of poverty, indebtedness, land alienation, displacement, deterioration of forest villages and the tribes living therein, shifting cultivation, etc. Solutions to this effect can best be found only when the deprivation and exploitation of tribes is eradicated. The Tenth Plan, therefore, adopted eradication of deprivation/exploitation of tribes as the centre-point in its approach, while pursuing simultaneously the Ninth Plan commitment of empowering the tribes.

12.4.12. Eleventh Five-Year Plan (2007-12)

The Eleventh Plan has experienced a paradigm shift with respect to the overall empowerment of the tribal people, keeping the issues related to governance at the centre. The operational imperatives of the Fifth Schedule, TSP 1976, Panchayat Ecologically Sensitive Area (ESA) 1996, Recognition of Forest Right Act (RFRA) 2006; the desirability of a tribalcentric, tribal-participative and tribal-managed development process; and the need for a conscious departure from dependence on a largely under-effective official delivery system will be kept in view during this shift.

12.4.13. Twelfth Five-Year Plan (2012-17)

Provision of 494.93 crores as SCA, 691.60 crores as Additional Central Assistance (ACA) under Article 275(1) of the Constitution and 900.00 crores as untied grant under the 'Janjati Kalyan Nidhi' for the development of tribal areas made in Twelfth Five-Year Plan (2012-17).

12.5. SCHEMES FOR TRIBAL DEVELOPMENT

The tribal majority areas in the country are broadly divided into three categories:

1. Predominantly tribal States and Union Territories
2. Scheduled area
3. Non-Scheduled areas in the States

All the tribal-majority States and Union Territories are placed in a special category for availing funds. The development and administration of tribal areas is accepted as a special responsibility of the Central Government even though they are integral parts of the concerned States. Financial provisions for their development were considered in detail by the constituent assembly itself.

The schemes have been divided into two categories:

1. Central sector programmes, which are fully financed by the Central Government.
2. Centrally-sponsored programmes, which are partly financed by the Central Government, and rest of the expenditure meted out by the concerned State Government.

According to Dr. B.D. Sharma, financial resources for developmental programmes in a State may comprise the following elements:

- Investment in the Central and Center-sponsored schemes
- State revenues
- Share from certain Central revenue
- Plan assistance from the Central Government
- Grants under Article 275 (1) on the basis of the recommendations of the Finance Commission

The utilization of State funds is broadly classified under two categories: Plan and non-plan. The plan technically covers all those items, which are included in the State or the Central plan. The non-plan includes expenditure on general administration as also on the maintenance of development schemes.

The SCA for TSPs is allocated between different States on the basis of three criteria as under:

1. The tribal population of Sub-plan area
2. The geographical area of the Sub-plan
3. The per capita gross output of the State

The weightage for these three elements has been fixed in a certain proportion.

While the first two criteria are simple, the quantum of assistance on the basis of the third criterion is determined with reference to the difference between the inverse of the State's per capita gross product and the inverse of the per capita gross national product.

The financing agencies rendering their services in the tribal areas are Central Government, State Governments, institutions including commercial banks, co-operative banks, NABARD and voluntary organizations.

12.6. PROGRAMMES FOR TRIBAL WELFARE

A number of employment-oriented and developmental programmes for tribals have been introduced by the Government of India. The major programmes are Integrated Rural Development Programme (IRDP), Jawahar Rojgar Yojana (JRY), Prime Ministers' Rojgar Yojana (PMRY) and Training of Rural Youth for Self Employment (TRYSEM). IRDP scheme is only for rural people that belong to below poverty line and others are for both rural as well as urban youth. All these schemes are implemented in the State by District Rural Development Agencies (DRDAs) in collaboration with Commercial and Co-operative Banks. PMRY was initiated in October 1993 to tackle the burning problem of educated unemployment. PMRY relates to setting up of self employment ventures through industries and services. Any unemployed youth who is matric failed/passed or above or ITI passed, is eligible for the benefits of the scheme subject to the condition that he is between the age group of 18-35 years and his family income does not exceed Rs. 24,000 per annum. The youth should also be the permanent resident of the area for at least three years and he should not be defaulter to any bank or financial institution. The scheme envisages 22.5% reservation for SC, ST and 27% for OBC. A maximum loan of Rs. 1 lakh per candidate is provided under this scheme, at an interest rate of 12.5-15.5%. The entrepreneur has to contribute 5% of project cost as margin money. No collateral security guarantee is asked on such loans. Period of repayment starts after a moratorium of 6-18 months and ranges over 3-7 years. The government provides subsidy to the extent of 15% of the total loan disbursed with a ceiling of Rs. 7,500 per entrepreneur. In case of a joint venture, each partner may be provided a loan of Rs. 1 lakh subsidy. In such cases, the interest is calculated for each partner separately at a rate of 15% of his share in the project cost limited to Rs. 7,500 for each partner.

The provision of compulsory training to entrepreneurship development is a salient feature of the scheme. Duration of this training is one month and trainees are provided stipend of Rs. 500 during the training period.

With the objective of providing technical skills to rural youth to enable them to take up self employment in the fields of agriculture and allied activities, industries, services and business activities, the scheme of TRYSEM was introduced in the year 1979 on the Independence Day. The scheme works as a part of IRDP and aims at imparting training to about two lakh rural youths every year from the 5,011 development blocks of the country, and to lift them above the poverty line.

Under this scheme, those rural youth, who are in the age group of 18-35 years with annual timely income from all sources not exceeding Rs. 8,500, are eligible for selection. Preference is given to SC and ST (50%), women (40%) and physically handicapped (3%). The selected beneficiaries are trained into the field of agriculture and allied activities, industry, service and business activities. The trainees are paid a monthly stipend or daily allowance during the training period. The training institution is provided honorarium along with Rs. 75 per trainee per month for purchase of raw materials.

After completion of training, the beneficiaries are assisted in getting finance from the banks. A maximum of Rs. 35,000 can be sanctioned to each beneficiary as a composite loan. Trained youths are granted a subsidy by the government at the rate of one-third to one-half of the cost of the project to set up self-employment venture. Beneficiaries are also provided marketing support for their finished products. The amount spent on this scheme is contributed by the Central and the State governments in 50:50 ratio. Since 1983, the scope of TRYSEM scheme has been enlarged.

The main thrust of the development strategies during the recent past has been on the removal of poverty in tribal areas. The main causes of poverty are identified as illiteracy, unemployment, under employment and low productivity in agriculture. Since farmers in TSP area have land holding mostly on hill slopes, the fertility of land is very low. Further, droughts and soil erosion are now recurring features in the tribal areas. This has reduced employment opportunities of the tribals. For improving the economic status of the tribals, special programmes were launched during 1980's, mainly the Asset Programme and the Employment Programme.

1. The Asset Programme aims at the overall integrated development of rural life through the removal of poverty and unemployment in rural areas. In this programme, productive assets are directly given to the poor. It is believed that income generated from these productive assets would not only be sufficient to repay the bank loans, but will help the assisted families to cross the 'poverty line'. This programme is popularly known as IRDP.
2. The Employment Programme, on the other hand, aims at providing employment through public works during the adverse agricultural season. The employment programme asserts that poverty persists because of the lack of employment opportunities. The earlier employment schemes were ad hoc in nature, but the employment programme launched from October 1980, popularly known as National Rural Employment Programme (NREP), is considered as a permanent plan programme.

The travails of tribal development need to be understood properly. The programmes should be related to the specific needs of the tribal community. Also, tribal development programmes should be integrated with the ongoing rural development programmes meant for poverty alleviation. A pragmatic and holistic approach to tribal development alone can produce good results.

12.7. NATIONAL SCHEDULED TRIBES FINANCE AND DEVELOPMENT CORPORATION (NSTFDC)

National Scheduled Tribes Finance and Development Corporation (NSTFDC) started in April, 2001 as a Government of India company under MoTA. It is managed by the Board of Directors with representation from Central Government, State Channelizing Agencies (SCAs), NABARD, IDBI, TRIFED and eminent persons from the STs.

12.7.1. Objectives and Functions of NSTFDC

NSTFDC is an Apex organization under MoTA for providing financial assistance for the economic development of the STs. The broad objectives and functions of NSTFDC are:

- To identify economic activities of importance to the STs so as to generate self-employment and raise their income level
- To upgrade their skills and processes through both institutional and on-the-job training
- To make existing State/Union Territory STFDCs and other developmental agencies engaged in economic development of the STs more effective.
- To assist SCAs in project formulation, implementation of NSTFDC-assisted schemes and in imparting training to their personnel
- To monitor implementation of NSTFDC-assisted schemes to assess their impact
- To generate awareness amongst the STs about NSTFDC concessional schemes
- To provide assistance for skill development and capacity building of the beneficiaries as well as of the officials of SCAs
- To provide concessional finance for viable income-generating schemes through SCAs and other agencies for economic development of the eligible STs
- To assist in market linkage of tribal produce

12.7.2. Schemes under Income-generating Activities

The Corporation provides financial assistance for income-generating activities and marketing support assistance for economic upliftment of STs. The details of schemes of NSTFDC are as under:

Term Loan Scheme

NSTFDC provides term loan for viable projects costing upto Rs. 10 lakh per unit. Under the scheme, financial assistance is extended upto 90% of the cost of the project and the balance is met by way of subsidy/ promoter's contribution/ margin money.

Adivasi Mahila Sashaktikaran Yojana (AMSY)

This is an exclusive scheme for economic development of ST women. Under the scheme, NSTFDC provides loan upto 90% for projects costing upto Rs. 50,000. Financial assistance under the scheme is extended at highly concessional interest rate of 4% per annum.

Micro Credit Scheme

This is an exclusive scheme for Self Help Groups (SHGs) for meeting small loan requirement of ST members. Under the scheme, the Corporation provides loans upto Rs. 35,000 per member and Rs. 5 lakh per SHG.

Scheme for SHGs

This is also a specific scheme for SHGs and financial assistance is extended for projects having unit cost upto Rs. 25 lakh per SHG. NSTFDC provides financial assistance upto 90% of the cost of the project subject to per member loan not exceeding Rs. 50,000. Minimum promoter contribution is 10% of the cost of the project.

Assistance to TRIFED Empanelled Artisans

The objective of the scheme is to extend concessional finance for purchase of project-related assets and working capital for tribal artisans empanelled with TRIFED. Under the scheme, the financial assistance is provided upto Rs. 50,000 for individuals and upto Rs. 5 lakh per SHGs/ Co-operative Societies.

Adivasi Shiksha Rrinn Yojana (Education Loan Scheme)

The objective of the scheme is to enable ST students to meet expenditure for pursuing professional/technical education including PhD in India. Financial assistance is provided upto Rs. 5 lakhs at concessional rate of interest of 6% per annum. Concurrence of Ministry of Human Resource Development, Government of India has also been obtained for providing interest subsidy to ST beneficiaries assisted under the scheme. ST students on completion of education can further avail loan for setting up income-generating units.

Marketing Support Assistance

The Corporation provides financial assistance to meet working capital requirement of agencies engaged in procurement and marketing of MFP and other tribal products.

12.8. TRIBAL CO-OPERATIVE MARKETING DEVELOPMENT FEDERATION OF INDIA LIMITED (TRIFED)

TRIFED is a Multi-State Co-operative Society. It was set up in 1987 under the Multi-State Co-operative Societies Act, 1984. TRIFED is now functioning both as a service provider and market developer for tribal products. Further, in the role as a capacity builder, it imparts training to ST artisans and, MFP, gatherers. Various schemes under TRIFED are discussed below:

12.8.1. Marketing Support

TRIFED is undertaking marketing of tribal products through the network of its retail outlets 'TRIBES INDIA' throughout the country. During the year 2012-13,

TRIFED has made sales of tribal products worth Rs. 932.36 lakh as on 31.12.2012; the details of marketing development activities has been as under:

TRIFED extended its chain of TRIBES INDIA outlets by adding six own outlets at New Delhi, Khajuraho (Madhya Pradesh), Kovalam (Kerala), Kanyakumari (Tamil Nadu), Hyderabad and Pinjore (Haryana), and one outlet on consignment basis at Guwahati. With these outlets, TRIFED has established 32 own TRIBES INDIA outlets and nine outlets on consignment basis marking its presence in 34 cities of 19 States. TRIFED also participates in exhibitions in India and abroad for marketing of tribal produce.

12.8.2. Tribal MFP Gatherers' Training and Capacity Building

TRIFED undertakes the following major projects/activities:

- Mahua Flower Development Activity: Regular Training programmes are organized on best practices of collection, drying, grading, storage and marketing of Mahua Flower in various States.
- Lac Cultivation and Processing Activity: On-farm trainings are provided on regular basis on lac collection and its processing.
- Honey Gatherers' Development Activity: Skill upgradation/capacity-building training on scientific collection/cultivation, processing, value addition, storage, marketing, etc.
- Hill Grass Training: Training programmes are organized to develop and upgrade the skills of tribals to produce value-added brooms based on market-driven products for increasing their livelihood.
- Training on Precision Agriculture Technology: Training on precision agricultural technology on medicinal herbs and other allied agricultural activities are organiszd in tribal areas.
- Dona Pattal Activity: Capacity building programmes are undertaken on Dona Pattal making activity by using moulding machines.

12.8.3. Tribal Artisans Training and Capacity Building

TRIFED is actively engaged in organization of training programmes on 22 crafts for beneficiaries in 17 States and one Union Terrotory viz., Jharkhand, Odisha, West Bengal, Gujarat, Maharashtra, Karnataka, Andhra Pradesh, Tamil Nadu, Sikkim, Assam, Nagaland, Meghalaya, Manipur, Tripura, Uttarakhand, Himachal Pradesh, Jammu and Kashmir, and Andaman and Nicobar Islands.

The categories of crafts introduced in the trainings are Bagh Print, Cotton Carpet, Wax Candle making, Jute craft, Toda craft, Soura Painting, Handloom weaving, Dokra craft, Warli painting, Banjara painting, Textile craft, Tribal painting, Bell metal, Wollen/ Pashmina, File folder (gift and assortment item), Pithora painting, Gota work, Handmade paper, Lac craft, Bamboo craft, Patch work, Embroidery, etc.

12.8.4. Research and Development Activity

TRIFED has been undertaking R & D activities by sponsoring research on number of NTFPs. In TRIFED's context, R&D is an important aspect as tribals from diverse income levels, age groups, and cultural backgrounds, harvest NTFPs for household subsistence, maintaining cultural and family traditions, obtaining spiritual fulfillment, maintaining physical and emotional well-being, scientific learning, and earning income. But, they still use traditional collection, processing techniques with limited applications, and market it as raw or with minimum processing. Most of the NTFPs are underutilized and there are few organizations, which are working on the MFPs in the country. It is in this background that research and development on NTFP assumes critical importance for enhancing the income of tribal collector gatherers, reducing drudgery and waste, and increasing efficiency of tribals. TRIFED, being the apex organization for providing marketing support for tribal products, is carrying on the task of sponsoring research and development projects with the financial support of MoTAs.

12.9. THE STS AND OTHER TRADITIONAL FOREST DWELLERS (RECOGNITION OF FOREST RIGHTS) ACT, 2006

Forest dwellers are among this country's poorest people. For many tribal people and other forest communities, forests are the source of livelihood, identity, customs and traditions. The forest dwelling STs and other traditional forest dwellers inhabiting forests for generations were in occupation of the forest land for centuries. However, their rights on their ancestral lands and their habitats had not been adequately recognized, despite them being integral to the very survival and sustainability of the forest ecosystem. The traditional rights and interests of forest dwelling STs and other traditional forest dwellers on forest lands were left unrecognized and unrecorded in the past through faulty reservation process during consolidation of State forests.

The forest-dwelling tribal people and the forests are inseparable, a factor that also ensures conservation of ecological resources stemming from the very ethos of tribal life. The conservation processes for creating wilderness and forest areas for production forestry, somehow ignored the bona fide interests of the tribal community from legislative framework in the regions where tribal communities primarily inhabited. The simplicity of tribals and their general ignorance of modern regulatory frameworks precluded them from asserting their genuine claims to resources in areas where they belonged and depended upon. The modern conservation approaches also advocated exclusion rather than integration. It was much later that forest management regimes initiated action to recognize the occupation and other rights of the forest dwellers, and integrated them in designs of management. Insecurity of tenure and fear of eviction from the lands where they had lived and thrived for generations were perhaps the biggest reasons why tribal communities felt emotionally as well as physically alienated from forests and forest lands. This historical injustice needed correction and, therefore, the Government enacted the Scheduled Tribes and Other Traditional Forest Dwellers (Recognition of Forest Rights) Act, 2006, which is commonly known as FRA. The Act came into operation with the notification of Rules on 01-01-2008 for carrying out the provisions of the Act.

The Act also goes beyond the 'recognition' of forest rights and also empowers the forest rights holders, Gram Sabhas and local-level institutions with the right to protect, regenerate, conserve and manage any community forest resource. This marks a decisive step forward in resource governance itself. Hailed rightly as a milestone in the history of tribal peoples' and forest dwellers' movements, the Act endeavours to facilitate their political empowerment to govern the forests for sustainable use and conservation. Precisely, for these reasons, it becomes important to take firm proactive steps to make the necessary paradigm shift, particularly by the State Governments.

12.9.1. Salient Features of the FRA-2006

Section 3 of the Act lists the rights, which shall be the forest rights of the forest-dwelling STs and other traditional forest dwellers. The basic provisions under this Act are given below:

a. Right to hold and live in the forest land under the individual or common occupation for habitation or for self-cultivation for livelihood by a member or members of a forest-dwelling ST or other traditional forest dweller.

b. Community rights such as nistar, by whatever name called, including those used in erstwhile princely States, *Zamindari* or such intermediary regimes.

c. Right of ownership, access to collect, use, and dispose of MFP, which has been traditionally collected within or outside village boundaries.

d. Other community rights of uses or entitlements such as fish and other products of water bodies, grazing (both settled or transhumant) and traditional seasonal resource access of nomadic or pastoralist communities.

e. Rights including community tenures of habitat and habitation for primitive tribal groups and pre-agricultural communities.

f. Rights in or over disputed lands under any nomenclature in any State where claims are disputed.

g. Rights for conversion of Pattas or leases or grants issued by any local authority or any State Government on forest lands to titles.

h. Rights of settlement and conversion of all forest villages, old habitation, unsurveyed villages and other villages in forests, whether recorded, notified or not into revenue villages.

i. Right to protect, regenerate or conserve or manage any community forest resource, which they have been traditionally protecting and conserving for sustainable use.

j. Rights, which are recognized under any State law or laws of any Autonomous District Council or Autonomous Regional Council or which are accepted as rights of tribals under any traditional or customary law of the concerned tribes of any State.

k. Right of access to biodiversity and community right to intellectual property and traditional knowledge related to biodiversity and cultural diversity.

l. Any other traditional right customarily enjoyed by the forest-dwelling STs or other traditional forest dwellers, as the case may be, which are not mentioned in clause (a) to (k), but excluding the traditional right of hunting or trapping or extracting a part of the body of any species of wild animal.

m. Right to in situ rehabilitation including alternative land in cases where the STs and other traditional forest dwellers have been illegally evicted or displaced from forest land of any description without receiving their legal entitlement to rehabilitation prior to the 13th day of December, 2005.

Section 3(2) of the Act provides for diversion (1) of forest land for certain facilities managed by the Government notwithstanding anything contained in the Forest Conservation Act, 1980 and, which involves felling of trees not exceeding 75 trees per hectare, and the forest land to be diverted for the purpose is less than one hectare, and the clearance of such development project is recommended by the Gram Sabha. Section 4(1) of the Act recognizes and vests (2) forest rights in forest-dwelling STs and other traditional forest dwellers notwithstanding anything contained in any other law for the time being in force.

Section 4(2) of the Act provides for (3) modification or resettlement of forest rights in critical wildlife habitats of National Parks and Sanctuaries for the purpose of creating inviolate areas for wildlife conservation subject to fulfilment of conditions such as process of recognition of rights is complete in all the areas under consideration, no other reasonable option exists, it has been established that the activities or the impact of the present right holders will cause irreversible damage and threaten the existence of wildlife and their habitat, free and informed consent of the concerned Gram Sabhas has been obtained, resettlement or alternative package has been prepared and communicated that provides a secure livelihood for the affected individuals and communities and fulfils the requirements of such affected families and communities given in the relevant laws and policy of the Central Government, and the resettlement should take place only after the facilities and land allocation at the resettlement location are complete. It is also provided that critical wildlife habitats, from which right holders are relocated, shall not be subsequently diverted for other users.

Section 4(3) of the Act subjects the recognition (4) and vesting of the forest rights to the condition that the forest-dwelling STs or other traditional forest dwellers had occupied the forest land before the 13th day of December, 2005. Section 4(4) stipulates that the rights conferred under this Act shall be heritable, but not alienable or transferable, and shall be registered jointly in the name of both the spouses in case of married persons.

A very important and crucial safeguard has (5) been provided to the forest right holders vide Section 4(5), which mandates that no forest-dwelling STs and other traditional forest dwellers shall be evicted or removed from the forest land under his occupation till the recognition and verification procedure is complete.

The right to hold and live in the forest land (6) under the individual or common occupation for habitation and for self-cultivation for livelihood by a member or members of a forest-dwelling STs or other traditional forest dwellers shall be restricted to area under actual occupation and shall in no case exceed an area of 4 hectares as per Section 4(6) of the Act. Section 4(7) of the Act provides that forest rights shall be conferred free from all encumbrances and procedural requirements.

The forest rights recognized and vested (7) under this Act includes the right of land to forest-dwelling STs and other traditional forest dwellers who can establish that they were displaced from their dwelling and cultivation without land compensation due to State development interventions and where the land has not been used for the purpose for which it was acquired within five years of the said acquisition as per Section 4(8) of the Act.

Section 5 of the Act empowers the holders (8) of forest rights, the Gram Sabha and village level institutions to (a) protect the wildlife, forest and biodiversity, (b) ensure the adjoining catchment area, water sources and other ecological sensitive areas are adequately protected (c) ensure that the habitat of forest-dwelling STs and other traditional forest dwellers is preserved from any form of destructive practices affecting their cultural and natural heritage and (d) ensure that the decisions taken in the Gram Sabha to regulate access to community forest resources and stop any activity, which adversely affects the wild animals and the bio-diversity are complied with.

Section 6 (Chapter IV) of the Act deals with (9) the authorities and procedures to vest forest rights in forest-dwelling STs and other traditional forest dwellers. There is a three-tier structure of authorities to vest forest right; the Gram Sabha is the initiating authority for determining the nature and extent of individual or community forest rights or both that may be given to the forest-dwelling STs and other traditional forest dwellers. The Sub-divisional Level Committee examines passed by the Gram Sabha and forwards it to the District-level Committee for the final decision. Any person aggrieved by the resolution of the Gram Sabha may prefer a petition to the Sub-divisional Level Committee and any person aggrieved by the decision of the Sub-divisional Level Committee may prefer a petition to the District-level Committee. The decision of the District-level Committee on the record of forest rights is final and binding. There is a State-level monitoring the resolution Committee to monitor the process of recognition and vesting of forest rights, and to submit such returns and reports to the nodal agency as may be called by that agency.

Section 7 of the Act has a provision for (10) punishment of any authority or officer for contravention of any provision of this Act or any rule made there under.

The MoTA, Government (11) of India is the nodal agency for the implementation of this Act as per Section 11 of the Act.

Section 12 empowers the Central Government to issue directions to the authorities referred to in Chapter IV of the Act.

As per Section 13 of the Act, the provisions (13) of this Act shall be in addition to and not in derogation of the provisions of any other law for the time being in force.

13

TRIBALS OF MAHARASHTRA

13.1. INTRODUCTION

Maharashtra occupies the Western and Central part of India, and has a long coastline stretching nearly 720 kms along the Arabian Sea. The Sahyadri mountain ranges provide a physical backbone to the State on the West, while the Satpuda hills along the North and Bhamragad–Chiroli–Gaikhuri ranges on the East serve as its natural borders. The State has 35 districts, divided into six revenue divisions for administrative purposes including Konkan, Pune, Nashik, Aurangabad, Amravati and Nagpur. Maharashtra is the second largest State in India in terms of geographical area, spread over 3.08 lakh sq km and in terms of population. The State's population, which is 9.29% of the entire country's population, is 11.24 crore. Majority of the forest area provides livelihood to tribals of Maharashtra State. Districtwise forest area is shown in Table 17.

13.2. TRIBAL DEMOGRAPHY

In Maharashtra, geographically, the concentration of tribals is confined in three traditional regions i.e., Sahyadri, Satpudas, and Gondwana region. The Sahyadri region comprises of Thane, Raigad, Nashik and Pune districts, where *Warlis*, *Kokanas*, *Thakurs*, *Mahadeo Kolis*, *Malhar Kolis*, *Dhor Kolis*, *Katkaris*, etc. tribal groups are found. The Satpudas region incorporates Nandurbar, Dhule, Jalgaon, Aurangabad and Amravati districts and particularly, *Bhils*, *Kokanas*, *Dublas*, *Dhankas*; *Korkus* are the main tribal groups here. The Gondwana region includes districts such as Chandrapur, Gadchiroli, Bhandara, Yavatmal and Nagpur, and *Gonds*, *Madia Gonds*, *Korkus*, *Kolams*, *Paradhans*, *Andhs*, etc. are the main tribal groups found in this region. List of tribes found in Maharashtra State is shown in Table 18.

Maharashtra supports livelihood for 45 tribes distributed in all the districts of the State (Fig. 31). *Katkaria* (*Kathodi*), *Maria Gond* and *Kolam* are among the PTGs of India and are mainly distributed in Yavatmal, Nanded, Osmanabad, Chandrapur and Gadchiroli districts of Maharashtra. Forest provides main livelihood support for the tribal economy in different parts of the State. Fig. 32 shows distribution of forest in Maharashtra State.

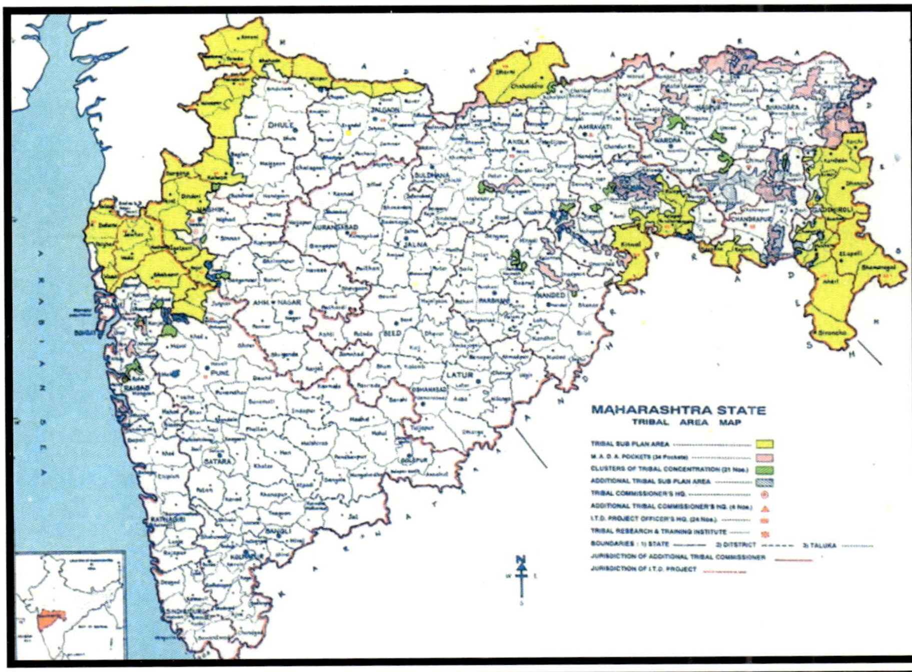

Fig. 31. Tribal area in Maharashtra State (Source: TRTI, Pune, 2015)

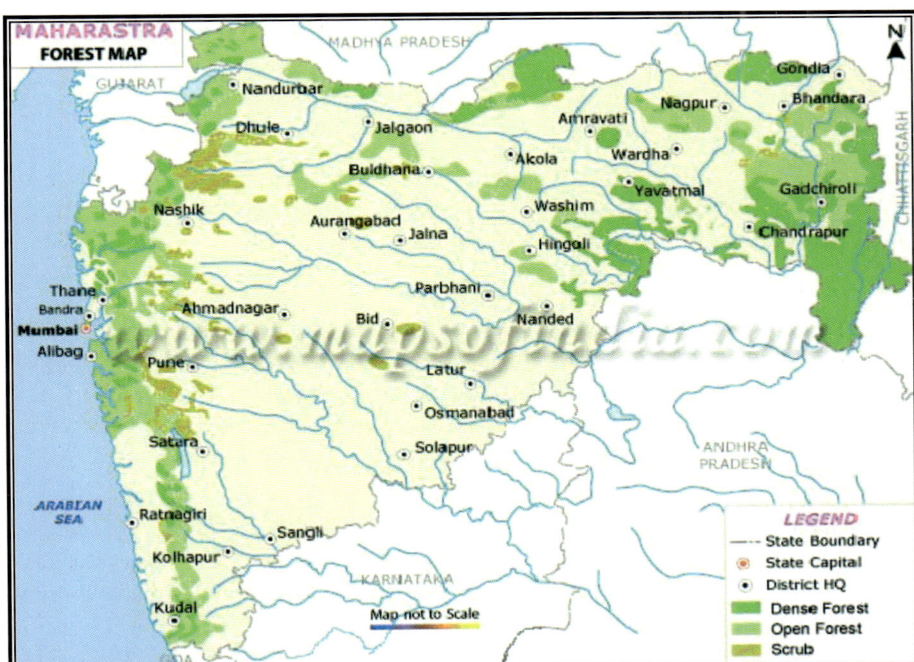

Fig. 32. Distribution of Forest in Maharashtra State
(Source: Maharashtra Forest Department, 2013)

Table 17: Forest cover of tribal districts in Maharashtra State (Area in sq km)

Sr. No.	District	Geographical Area	Very Dense	Moderately Dense	Open Dense	Total	Scrub
1.	Ahmadnagar	17048	0	69	217	286	555
2.	Amaravati	12210	655	1455	1077	3187	116
3.	Chandrapur	11443	1340	1588	1150	4078	56
4.	Dhule	7189	0	70	251	321	103
5.	Gadchiroli	14412	4733	3392	1969	10094	20
6.	Jalgaon	11765	52	363	770	1185	69
7.	Nagpur	9892	372	953	698	2023	77
8.	Nashik	15530	0	351	738	1089	319
9.	Pune	15643	0	757	975	1732	493
10.	Thane	9558	0	1281	1631	2912	222
11.	Yavatmal	13582	123	1110	1372	2605	97
	Total	138272	7275	11389	10848	29512	2127

Source: IFSR-2011

Table 18: List of tribes of Maharashtra

Sr. No.	Tribes of Maharashtra
1.	*Andh*
2.	*Baiga*
3.	*Barda*
4.	*Bavacha, Bamcha*
5.	*Bhaina*
6.	*Bharia Bhumia, Bhuinhar Bhumia, Pando*
7.	*Bhattra*
8.	*Bhil, Bhil garasia, Dholi, Bhil, Dangri Bhil, Dungri Garasia, Mewsi Bhil, Rawal, Bhil, Tadvi Bhil, Bhagalia, Bhilala Pawra, Vasava, Vasave*
9.	*Bhunjia*
10.	*Binjhwar*
11.	*Birhul, Birhor*
12.	*Chodhara* (excluding Akola, Amravati, Bhandara, Buldana, Chandrapur, Nagpur, Wardha, Yavatmal, Aurangabad, Beed, Nanded, Osmanabad and Parbhani districts)
13.	*Dhanka, Tadvi, Tetaria, Valvi*
14.	*Dhanwar*
15.	*Dhodia*

Sr. No.	Tribes of Maharashtra
16.	*Dubla Talavia, Halpati*
17.	*Gamit, Gamta, Gavit, Mavchi, Padvi*
18.	*Gond, Rajgond, Arakh, Arrakh, Agaria, Asur, Bedi Maria, Bada Maria, Bhatola, Bhimma, Bhuta, Koilabhuta, Koilabhuti, Bhar, Bisonhorn Maria, Chota Maria, Dandami Maria, Dhuru, Dhurwa, Dhoba, Dhulia, Dorla, Kaiki, Gatta, Gatti, Gaita, Gond Gowari, Hill Maria, Kandara Kalanga, Khatola, Koitar, Koya, Khirwar, Khirwara, Kucha Maria, Kuchaki, Maria, Media, Maria, Mana, Meannewar, Moghya, Mogia Moghya, Mudia, Muria, Nagarchi, Naikpod, Nagwanshi, Ojha, Raj Sonjhari Jhareka, Thatia, Thotya, Wade Maria, Vade, Maria.*
19.	*Halba, Balbi*
20.	*Kamar*
21.	*Kathodi, Katkari, Dhor Kathodi, Dhor Kathkari, Son Kathodi, Son Katkari*
22.	*Kawar, Kanwar, Kaur, Cherwa, Rathia, Tanwar, Chattri*
23.	*Khairwar*
24.	*Kharia*
25.	*Kokna, Kokni, Kukna*
26.	*Kol*
27.	*Kolam, Mannervarlu*
28.	*Koli dhor; Tokre Koli, Kolcha, Kolgha*
29.	*Koli Mahadev, Dongar Koli*
30.	*Koli Malhar*
31.	*Kondh, Khond, Kandh*
32.	*Korku, Bopchi, Mouasi, Nihal, Nahul, Bondhi, Bondeya*
33.	*Koya, Bhine Koya, Rajkoya*
34.	*Nagesia, Nagasia*
35.	*Naikda, Nayaka, Cholivala Nayaka, Kapadia Nayaka,, Mota Nayaka, Nana Nayaka*
36.	*Oraon, Dhangad*
37.	*Pardhan, Pathari, saroti*
38.	*Pardhi, Advichincher, Phans Pardhi, Phanse Pardhi, Langoli Pardhi, Behelia, Behellia, Chita Pardhi, Shikari, takankar, Takia*
39.	*Parja*
40.	*Patelia*
41.	*Pomla*
42.	*Rathwa*
43.	*Sawar, Sawara*
44.	*Thakur, Thakar, Ka Thakar, Ma Thakur, Ma Thakar*

Sr. No.	Tribes of Maharashtra
45.	*Thoti* (in Aurangabad, Beed, Nanded, Osmanabad and Parbhani districts and Rajura tahsil of Chandrapur District)
46.	*Warli*
47.	*Vitolia, Kotwalia, Barodia*
48.	*Vitolia, Kotwalia, Barodia*

Source: Annual Report, MoTA, Government of India, 2012-13

Table 19: Districtwise dominant tribes of Maharashtra State

Sr. No.	District	Major Tribes
1.	Thane	*Kathodi, Katkari, Kokna, Kokni, Koli, Mahadeo, Koli Malhar, Warli, Thakar, Dubla, Koli, Dhor, Tokre Koli*
2.	Raigad	*Kathodi. Koli Mahadeo, Katkari*
3.	Ratnagiri	*Kathodi. Koli Mahadeo, Katkari, Dongar Koli*
4.	Sindhudurg	*Kathodi. Koli Mahadeo, Katkari, Dongar, Koli*
5.	Nashik	*Bhil, Kokana, Koli Mahadeo, Thakar, Warli, Kathodi, Katkari*
6.	Dhule	*Bhil, Gamit, Kokana, Koli Dhor, Dhanka, Tokare Koli, Naikada, Pardhi, Vitolia*
7.	Jalgaon	*Bhil, Koli Dhor, Dhanka, Pardhi, Tokare Koli*
8.	Ahmednagar	*Bhil, Koli Mahadeo, Thakar, Koli Dhor, Tokare Koli, Pardhi*
9.	Pune	*Koli Mahadeo, Thakar, Kathodi, Katkari, Koli Dhor, Tokare Koli*
10.	Satara	*Koli Mahadeo, Kathodi, Katkari*
11.	Sangli	*Koli Mahadeo*
12.	Solapur	*Koli Mahadeo, Pardhi*
13.	Kolhapur	*Koli Mahadeo*
14.	Aurangabad	*Bhil, Koli Mahadeo, Koli Malhar*
15.	Jalna	*Bhil, Koli Mahadeo, Koli Malhar*
16.	Parbhani	*Andh, Koli Mahadeo*
17.	Beed	*Bhil, Koli Mahadeo*
18.	Nanded	*Andh, Gond, Kolam, Koli Mahadeo, Oraon, Dhangad, Pardhan*
19.	Osmanabad	*Koli Mahadeo, Oraon, Dhangad*
20.	Latur	*Koli Mahadeo, Oraon, Dhangad*
21.	Buldhana	*Andh, Dhanwar, Koli Mahadeo, Pardhi*
22.	Akola	*Andh, Dhanwar, Koli Mahadeo, Pardhi*
23.	Amravati	*Dhanwar, Gond, Korrku, Pardhi, Nhihal*
24.	Yavatmal	*Andh, Dhanwar, Gond, Kolam, Pardhan, Halba, Halbi, Kharia, Pardhi*

Sr. No.	District	Major Tribes
25.	Wardha	*Gond, Pardhan*
26.	Nagpur	*Gond, Halba*
27.	Bhandara	*Gond, Halba*
28.	Chandrapur	*Gond, Halba, Pardhan, Halbi, Kawar, Kolam*
29.	Gadchiroli	*Gond, Halba, Pardhan, Halbi, Kawar, Kolam*
30.	Greater Mumbai	*Warli, Thakar, Dubla, Dhodia, Gond, Koli Dhor, Tokare Koli*

Source: Census of India, 1981; Maharashtra State Series No. 12, Part - IIB.

As per the Census of India-2011, there are 1,05,10,213 reported as ST, of which 53,15,025 are males and 51,95,188 are females showing an addition of 19,32,937 during 2001-11. Maharashtra is second largest tribal-populated State (next to Madhya Pradesh) contributing about 10.1% of total Indian tribal population. At district level, Nandurbar has the highest proportion of ST population with 69.3% followed by Gadchiroli (38.7%), Dhule (31.6%) and Nashik (25.6%). Lowest population of tribes found in Sangli district of Maharashtra. Districtwise dominant tribes of Maharashtra State are given in Table No. 19.

In Maharashtra State, sex ratio is at 925, which is lesser than the National average of 940 (Census, 2011). The sex ratio of the ST population in Maharashtra is 973 females per 1,000 males, which is higher than 922 reported for the State population as a whole in 2001.

LR in Maharashtra has seen upward trend and is 82.34% as per 2011 Population Census (India 74.04%). According to 2011 Census, percentage of literate rate among ST population of Maharashtra is 65.7% (ST of India is 59.0%).

13.3. MAHARASHTRA GOVERNMENT INITIATIVES FOR TRIBAL DEVELOPMENT

The following schemes and programmes are implemented by the Government of Maharashtra for overall development of tribals of Maharashtra:

- Ashram School (government and aided)
- Hostels – boys and girls (government and aided)
- Eklavya English medium schools
- Government of India – Scholarship Scheme
- Employment Registration – ST students
- Financial assistance for professional studies
- Cash awards to meritorious student
- Incentives to ST girl students to reduce dropout ratio
- Scholarship to ST – handicap student studying in 8th-12th standard

- Cash award to Ashram Schools
- Electric Motor Pump and Oil Engine Scheme
- Supply Gas Burner and suraksha tube
- Motor Driving Training Scheme
- Para-military and Police recruitment training
- Kanyadan Scheme
- Mahila Sabalikaran – women SHG
- Swabhiman Scheme
- Thakkar Bappa Scheme
- Jan Utkarsh Programme
- Nav Sanjeevan Yojana
- Consumption finance scheme
- Melghat Pattern – special health scheme
- Adiwasi Utthan Scheme – comprehensive rural health project
- Special Action Plan – for inaccessible tribal area
- Janashri Vima Yojana – group insurance scheme for tribals
- LPG on subsidized rate
- Various schemes under Article 275(1) of the Constitution of India – Central Government Scheme

ANNEXURE I

Table 20: Indian tribes and their major characteristics

S.N.	Name of Tribe	Distribution and Major Characteristics
1.	*Abhor*	These tribes are found in Arunachal Pradesh and Assam.
2.	*Abujmaria*	Known variously as *Abudjamadis*, *Abujmariya* and *Hill Maria*, these tribes are found in the geographically-inaccessible areas of Abujhmar Mountains and Kutrumar Hills in the Bastar district of Madhya Pradesh. They speak a Dravidian language called *Abujmaria*. The *Hill Maria* tribes are considered as a sub-group of the *Gonds*, who are historically the most important group of original Indian tribes.
3.	*Adivasika*	These are forest dwellers found mainly in Northern Kerala, near Calicut.
4.	*Adivasi Girasia*	These tribes inhabit the Banaskantha and Sabarkantha districts of Gujarat and are believed to be the descendants of the Rajputs who married *Bhil* women. The name '*Girasia*' refers to the Rajput and other landholders living in the Gujarat and Rajasthan regions. Their language, also known as *Adivasi Girasia*, is an Indo–Aryan language belonging to the *Bhil* sub-group.
5.	*Adiyan*	Also known as *Eravas*, these people are found in Kerala, Tamil Nadu and Karnataka.
6.	*Ahirs*	These people are the traditional dairymen caste of Uttar Pradesh.
7.	*Aka*	These tribes are found mainly in the Andaman Islands, Arunachal Pradesh and also in parts of Assam. The *Aka* people are so named for a black, sticky paint they use on their faces. They used to speak *Aka* (now an extinct language) on the Andaman Islands and *Aka Lel*, a dialect of *Nisi*, in Assam. The *Aka* people in Assam celebrate the Nechido festival every year on the first day of November.
8.	*Alars*	Also known as *Chathans* or *Chatans*, these tribes are found in the Kerala–Palghat region. They speak *Alar* and *Malayalam*.
9.	*Amindivi*	These tribes are found in Lakshadweep.
10.	*Amri Karib*	Known by different names such as *Mikir, Manchati, Mikiri, Karbi*; these tribes are found in the Mikir and Rengma Hills of Assam. They speak *Amri*, a dialect of *Mikir*.
11.	*Anal*	These tribes are found in Manipur.

S.N.	Name of Tribe	Distribution and Major Characteristics
12.	*Angami*	These tribes are found in Kohima; the capital of Nagaland.
13.	*Ao*	The *Ao* tribe is found in the Mokokchung District of Nagaland. The main festivals of the *Aos* are *Moatsu* and *Tsungremmong* celebrated during the first week of May and August, respectively.
14.	*Apatani*	These tribes, also described as *Apa*, are found in South of the Tibetan border in the States of Assam, Arunachal Pradesh and Nagaland. Their language is also known as *Apatani*. These tribes are renowned for their cultivation, especially the terraced-rice fields, which are located along the sides of the valleys.
15.	*Arnatas*	Also known as *Aranadan* and *Eranadans*, these tribes are found in Tamil Nadu, Karnataka and Kerala. They speak *Aranatan* and *Malayalam*.
16.	*Asurs*	These tribes are found in Bihar and West Bengal.
17.	*Awadhi*	Known by different names such as *Abadi, Abohi, Ambodhi, Baiswari, Kojali* and *Kosali*, these people are found in Bihar, Madhya Pradesh and parts of Uttar Pradesh. They speak in *Awadhi*, which is a dialect of *Hindi*.
18.	*Badaga*	Also known as *Badag, Badagu, Badugu* and *Vadagu*. These tribes are found in the Nilgiri and Kunda Hills of Tamil Nadu. Their language is also called *Badaga*. The name 'Badaga', meaning 'northerner', was given to this group during the Middle Ages when they migrated from the Mysore plains to the Nilgiri Hills in Southern Tamil Nadu.
19.	*Bagri*	These tribes are found mainly in Western India, in the States of Rajasthan, Punjab, Haryana, and Madhya Pradesh. They are Indo–Aryan people, and their language is also called *Bagri*.
20.	*Bajania*	These tribes are now found in Kachi Koli in Pakistan and are known by other names such as *Kachi Koli, Kuchikoli, Vaghri* and *Vagari*. They speak *Bajania*, a dialect of *Koli*.
21.	*Bakarwals*	These are the nomadic tribes of Kashmir.
22.	*Baiga*	Known by names such as *Baigai*, *Bega* and *Bhumia*, these people are found in Bihar, Maharashtra, Madhya Pradesh, Odisha and West Bengal. Their language is also known as *Baiga*.
23.	*Banchharas*	These tribes are found in Madhya Pradesh.
24.	*Bangni*	The *Bangni* (also known as the *Dafla* or the *Nishi*) inhabit the hills of Assam and Arunachal Pradesh. Their native language, *Nisi*, belongs to the Tibeto–Burman language family.
25.	*Bangri*	These tribes are located mainly in the States of Haryana, Karnataka, Himachal Pradesh, Uttar Pradesh, Punjab and Delhi. Their language, *Bangaru*, is a member of the Indo–Aryan language family.
26.	*Banjaras*	Known by different names in different places such as *Lamani*, *Lambadi*, *Bangala*, *Banjori*, *Banjuri*, *Gohar–Herkeri*, *Goola*, *Gurmarti*, *Kora*, *Labhani Muka*, *Lambara*, *Lavani*, *Lemadi*, *Lumadale*, *Sugali*, *Tanda*, *Vanjari*, *Waji*, *Gormati* and *Singali*, these tribes are mainly concentrated in

S.N.	Name of Tribe	Distribution and Major Characteristics
		Andhra Pradesh, Bihar, Madhya Pradesh, Himachal Pradesh, Gujarat, Tamil Nadu, Maharashtra, Karnataka, Odisha and West Bengal. Their common language is *Lamani*.
27.	*Barda*	These tribes are found in Gujarat.
28.	*Bavacha*	These tribes are found in Gujarat.
29.	*Barel*	The *Barels* are considered to be the sub-group of *Bhils*. They speak *Barel* language.
30.	*Bathudi*	The *Bathudis* live primarily in the districts of Mayurbhanj, Kendujhargarh and Balasore in the State of Odisha. Most *Bathudis* are bilingual, speaking their native language, *Bathudi* and *Oriya*.
31.	*Bauria*	The *Baurias* are also considered as a sub-group of *Bhils*. Their language is also known as *Bauria*.
32.	*Beda*	These tribes are found in Ladakh.
33.	*Bedia*	These tribes are found in Madhya Pradesh.
34.	*Bettakurubas*	These tribes are found in Karnataka.
35.	*Bhadrawahi*	These tribes are located predominantly in Jammu and Kashmir.
36.	*Bhils*	The *Bhils* are considered as the third largest and most widely-distributed tribal groups in India. The name '*Bhil*' was probably derived from the word villu or billu, which in most *Dravidian* languages is the word for 'bow'. The bow has long been a characteristic weapon of the *Bhil* because the tribesmen always carry their bows and arrows with them. The *Bhil* tribes inhabit some of the most remote and inaccessible areas of India. There are two divisions of *Bhils*: the Central or 'pure' *Bhils*, and the Eastern or Rajput *Bhils*. The Central *Bhils* live in the mountain regions in the States of Madhya Pradesh, Gujarat and Rajasthan. They are known as the connecting link between the Gujaratis and the Rajasthanis and are one of the largest tribal communities of India. They speak *Bhili*, which is an Indo–Aryan language. The *Bhils* are known to have fought against the Mughals, Marathas and the British.
37.	*Bhilala*	The *Bhilala* are located in several States in Western Central Indian, but mainly in the districts of Dhar, Jhabua, and West Nimar of Madhya Pradesh. Their language, also called *Bhilala,* is a sub-group of the *Bhil* language, which belongs to the Indo–Aryan linguistic family. The *Bhilala* are considered as nobility among the *Bhil*, since they are the direct descendants of the Rajput chiefs who took the daughters of the *Bhil* chieftains to be their wives.
38.	*Bhim*	These are the tribes found mainly in the Eastern State of Tripura.
39.	*Bhinjwari*	The *Binjhwaris* or the *Binjhals*, live in Madhya Pradesh, Maharashtra, Bihar and West Bengal. Although their native language is *Binjhwari*, many of these people also speak *Chhattisgarhi*.
40.	*Bhotia*	Known by names such as *Lahuli*, *Tinan*, *Lahauli*, *Rangloi* and *Gondla*, these tribes are found in Himalayan Mountain ranges of Himachal Pradesh. They speak in *Lahuli* and *Tinan*.

S.N.	Name of Tribe	Distribution and Major Characteristics
41.	*Bhutia*	These are Buddhist tribes found in Assam, along the border with Tibet and in Nepal. They are also known by the names *Lhasa*, *Dalai*, *Pohbetian* and *Tebilian*. They speak *Tibetan*.
42.	*Birhor*	These tribes are found in the Hazaribagh, Singbhum and Ranchi districts of Bihar and parts of Madhya Pradesh, Maharashtra, Odisha and West Bengal.
43.	*Bison Horn Maria*	These are small tribal groups located mainly in Gadchiroli district in Maharashtra and also in parts of Madhya Pradesh. The designation '*Bison Horn Maria*' is derived from a distinctive head-dress worn at marriage dances and adorned with the horns of now extinct wild bison. Many scholars believe that the *Bison Horn Maria* is part of the *Gond* tribe, while others consider them to be an earlier indigenous tribe assimilated by the *Gonds*.
44.	*Bombas*	These are the Nomads inhabiting the rugged hills of Kashmir.
45.	*Boro (Bodo)*	These tribes are found in Assam.
46.	*Braj Bhakha*	These tribes are located mainly in the States of Bihar, Madhya Pradesh and Uttar Pradesh. Their language, *Braj* Bhasha, is a member of the Indo–Aryan language family.
47.	*Bunde Soligas*	These tribes are found in Karnataka.
48.	*Chakmas*	Also known as *Takam*, these tribes are found in Assam, Meghalaya, Tripura, West Bengal and parts of Bangladesh. They speak *Chakma* language.
49.	*Chamars*	These tribes are known by different names such as *Chamari*, *Chambhar Boli* and *Chambhari*. These are found in the States of Madhya Pradesh, Uttar Pradesh and Maharashtra. Their language is known as *Chamari*.
50.	*Chameali Pahari*	The *Chameali Pahari* tribes are primarily located in the States of Uttar Pradesh, and Jammu and Kashmir.
51.	*Changs*	These tribes are found in the Tuensang district of Nagaland.
52.	*Chang–Pas*	These are the tribes found in the northern upland valleys of the Indus river in Jammu and Kashmir. They speak in *Tibetan* dialect.
53.	*Charan*	These tribes are found in Gujarat.
54.	*Chekhasang*	*Chekhasang* and *Pouchry* Tribes are found in the Phek district of Nagaland. *Chakhesang*'s culture and customs are quite different from the other *Naga* tribes. There are evidences of the existence of head-hunting among the villagers in the ancient days.
55.	*Chenchus*	Known variously as *Chenchucoolam*, *Chenchwar*, *Chenswar* and *Choncharu*, these tribes are found primarily in the State of Andhra Pradesh and also in parts of Tamil Nadu, Karnataka and Odisha. Their native language (also called *Chenchus*) belongs to the Dravidian language family. Many also speak *Telugu*.
56.	*Cheros*	These tribes, which claim to be the descendants of the Rajputs, are found in Bihar, Odisha, Uttar Pradesh and West Bengal and are primarily concentrated in Palamau, Shahabad, Champaran, and other surrounding districts. The *Cheros* speak a language that is also called *Chero*.

S.N.	Name of Tribe	Distribution and Major Characteristics
57.	*Chettier*	*Chettier* is the Hindu fishermen caste of Tamil Nadu.
58.	*Chola Naickans*	Also known as *Chola Naikar*, these tribes are found mainly in the Nilambur jungle in Kerala. They speak *Canarese*, a dialect of *Kannada*.
59.	*Dafla*	These tribes are found in Arunachal Pradesh.
60.	*Damarias*	These tribes are found in Rajasthan.
61.	*Deori*	These tribes live along the Brahmaputra river and are primarily located in the States of Assam, Nagaland and Arunachal Pradesh. They speak a Tibeto–Burman language that is also called *Deori*. The *Deoris* are one of the four divisions of the Chutiya people and are also related to the Eastern *Bodo–Garo*.
62.	*Dhodia*	The *Dhodia* are located in the extreme Southeastern districts of Gujarat, in the hilly regions South of the Tapi river, and in Dadra and Nagar Haveli. The *Dhodia* are the highest-ranking tribe and the third largest tribal group in Gujarat. They speak *Dhodia*, a *Bhil* language.
63.	*Dhurwa*	These tribes are forest dwellers found mainly in the Bastar district of Madhya Pradesh and Koraput district of Odisha. They are considered as a sub-group of the *Gond*, the largest tribal group in India. They speak *Parji* in three dialects.
64.	*Dimasa*	These are Proto–Austroloid tribes found in Meghalaya and Mizoram.
65.	*Dog–Pa*	Also known as *Srin*, *Shrin* and *Brog–Pa*, these tribes are found in Jammu and Kashmir. Their language is known as *Shrin*.
66.	*Dogris*	These tribes are known by various names such as *Dogri–Kangri*, *Dhogaryali*, *Dogari*, *Dogri Jammu*, *Dogri Pahari*, *Tokkaru* and *Dogri–Kangra*. They are mainly concentrated in Jammu and Kashmir between the Ravi and Chenab rivers. Many *Dogris* also live in Pakistan. The *Dogris* are a hardy people, divided into several castes and sects. Their language is known as *Dogri–Kangri*.
67.	*Dorli*	The *Dorli*, a sub-group of the *Gonds*, are concentrated in the Bastar district of Madhya Pradesh. Some also live in parts of Andhra Pradesh and Odisha. Their language is also called *Dorli*.
68.	*Dubla*	The *Dubla* live primarily in the States of Gujarat, Maharashtra, Karnataka, and Rajasthan. They speak *Dubla*, a *Bhil* language that belongs to the Indo–Aryan linguistic family. The *Dubla* tribe contains 20 sub-groups, of which the *Talavias* have the highest social rank.
69.	*Eravallan*	These tribes are found in Kerala.
70.	*Gaddi*	Known by names such as *Bharmauri Bhadi*, *Pahari Bharmauri*, *Panchi Brahmauri Rajput*, *Gaddyal* and *Gadi*, these tribes are found in parts of Himachal Pradesh, Madhya Pradesh, Uttar Pradesh and Punjab. Their language is also known as *Gaddi*.
71.	*Galong*	These tribes, also known as *Gallong*, *Gallo*, *Galo* and *Adi–Galo*, live in Assam, along the Tibet border area. Their language is also known as *Galong*.

S.N.	Name of Tribe	Distribution and Major Characteristics
72.	*Gamit*	These tribes are found in Karnataka.
73.	*Gamti*	This is one of the *Bhil* tribes that live mainly in the Surat and Bharuch districts of Gujarat, India. Among the *Bhil*, the word gamta means 'headman', possibly giving the *Gamti* a feeling of superiority over other *Bhil* tribes. They speak *Gamti*, which is one of the *Bhil* languages.
74.	*Garhwali*	The *Garhwali* or the Central Pahari are a hardworking and often isolated people who are primarily located in the States of Uttar Pradesh and Jammu and Kashmir.
75.	*Garos*	The *Garos* or *Achiks* belong to *Bodo* family of the Tibeto–Burman race and are found in Meghalaya. They are said to have migrated from Tibet.
76.	*Gonds*	The *Gonds* comprise the largest tribal group in India. Historically, the *Gonds* were the most important group of the original Indian tribes. In the 1500's, several *Gond* dynasties were established and their rajas or kings ruled like Hindu princes. The *Gonds* were conquered by the Muslim armies in 1592, but their tribes were not disturbed by the changes in administration.
77.	*Gongte*	These tribes are found in Manipur.
78.	*Gosains*	These tribes are found in Madhya Pradesh.
79.	*Gotte*	These tribes, also known as *Podia Koya*, are found in the jungles of Madhya Pradesh. Their language is *Podia Koya*, which is a dialect of *Koya*.
80.	*Gracias*	These tribes, known by different names such as *Garasia*, *Rajput Girasia*, *Dungri Grasia* and *Dhungri Bhili* are found in Gujarat and Rajasthan. Their language is known as *Garasia*.
81.	*Gujjars*	Known by names such as *Gujuri*, *Gujer*, *Gojri*, *Kashmir Gujari* and *Rajasthani Gujuri*, these are semi-nomadic tribal people found in Jammu and Kashmir, Himachal Pradesh, Madhya Pradesh, Uttar Pradesh and Rajasthan, besides parts of Afghanistan and Pakistan. Their language is known as *Gujuri* (also called *Parimu* and *Hindki*). Some historians believe that the *Gujjars* were the inhabitants of Georgia (*Gurjia*) a territory situated between the Black Sea and the Caspian Sea in the former Soviet Union. They left that area and migrated through Central Asia, Iraq, Iran and Afghanistan, crossed the Khyber Pass and entered the Indian Subcontinent. Several settlements are named after them e.g., *Gujar* (Central Asia), *Juzrs* (Gurjara), *Gujrabad*, *Gujru*, *Gujristan*, *Gujrabas*, *Gujdar–Kotta*, *Gujar–Garh*, *Gujarkhan* and *Gujranwala* in Iran and Afghanistan.
82.	*Halam*	*Halam* or *Malsum* tribes are found in Tripura. They originally hailed from one of the branches of *Kukis*.
83.	*Hanjis*	These tribes are found along the Jhelum river in the Kashmir valley.
84.	*Harauti*	The *Harautis* are located mainly in the Kota area of Rajasthan and also in parts of Madhya Pradesh.
85.	*Hilpulayan*	These tribes are found in Kerala.

S.N.	Name of Tribe	Distribution and Major Characteristics
86.	*Hmar*	These tribes are found in Mizoram.
87.	*Ho*	The word '*ho*' means man. These tribes, which are also known as *Lanka Kol* and Bihar *Ho*, are found mainly in the Singbhum district of Bihar and the Mayurbhanj district of Odisha, besides parts of West Bengal, Bangladesh and Nepal. Their language is also called *Ho*. These are classified as Caucasian.
88.	*Irula*	In the *Tamil*, the name *Irula* means 'people of darkness'. These tribes are known by different names such as *Eravallan, Erukala, Irava, Irular, Irular Mozhi, Iruliga, Iruligar, Korava, Yerukula* and *Kad Chensu*. They are mainly concentrated in Chingleput at the Nilgiri foothills in Tamil Nadu. They are also found in parts of Andhra Pradesh, Kerala, Tamil Nadu, Maharashtra and Karnataka. Their language, *Irula,* is related to *Tamil* and *Kannada.*
89.	*Jarawa*	(Also known as the *Onges*) These tribes are found on the Rutland Island and the South Andaman Island. They speak the *Jarawa* language.
90.	*Jatapu*	The *Jatapu* are mainly concentrated in the Srikakulam and Vishakhapatnam districts of Andhra Pradesh. Some communities are also found in the States of Tamil Nadu and Odisha. Their original language is called *Jatapu*, but most of the *Jatapu* speakers have adopted *Telugu* as their mother tongue.
91	*Jats*	These people form the dominant group in Haryana.
92.	*Juangs*	These tribes are found in the Banspal taluka of the Keonjhar district of Odisha.
93.	*Jaunsaris*	These tribes are found in Uttar Pradesh.
94.	*Kaikadi*	The *Kaikadis* are small tribal groups located mainly in Maharashtra and Karnataka. Their language (also called *Kaikadi*) is a member of the Dravidian language family.
95.	*Kamar*	These are the STs found primarily in the Raipur and Rewa districts of Madhya Pradesh. Their native language is also called *Kamar*.
96.	*Karbi*	These tribes are found in Assam.
97.	*Kashtwari*	They are located mainly in Jammu and Kashmir. Their language is known as *Kishtwari*.
98.	*Katkari*	They are located mainly in the States of Gujarat and Maharashtra. Their language is also known as *Katkari*.
99.	*Kathodia*	These tribes are found in Dadra and Nagar Haveli.
100.	*Kattunaickans*	These tribes are found in Kerala.
101.	*Kawar*	The *Kawars* are primarily located in the States of Madhya Pradesh, Odisha and Maharashtra. Their language, *Kawari*, is an Indo–Aryan language that is believed to be a dialect of *Halbi*.
102.	*Khambas*	These tribes are found in Arunachal Pradesh.
103.	*Khamtis*	Known by various names such as *Kham–Tai, Hkamti, Khampti, Khamti Shan, Khantis* and *Tai Kam Ti*. These tribes are found in Assam, Arunachal Pradesh and parts of Myanmar. Their language is described as *Khamti*.

S.N.	Name of Tribe	Distribution and Major Characteristics
104.	*Khandeshi*	They are located mainly in the States of Maharashtra and Gujarat. Their language is also known as *Khandeshi*.
105.	*Khasis*	These tribes are found in Khasi–Jaintia hills in Meghalaya, Manipur, Jammu and Kashmir, Punjab, Uttar Pradesh and parts of West Bengal. They are also described by other names such as *Khasi Paharis*, *Khuchia*, *Kassi*, *Khasa* and *Khashi*. Their language is *Khasi*. *Khasis* follow the matrilineal system of lineage.
106.	*Khiamungans*	These tribes are found in the Tuensang district of Nagaland.
107.	*Khirwar*	The *Khirwars*, a sub-group of the *Gonds*, live in the Sarguja district at the borders of Madhya Pradesh and Uttar Pradesh. Some speak the local *Hindi* dialect in addition to their own language, *Khirwari*.
108.	*Khorkhas*	These tribes are found in Jammu and Kashmir.
109.	*Kipgen*	These tribes are found in Manipur.
110.	*Kiutze*	These tribes are located mostly near the Myanmar–Tibet border. Their language is known as *Rawang*.
111.	*Koch*	These tribes live primarily in the States of Assam and Tripura. Their language, *Koch*, belongs to the Tibeto–Burman language family. In the sixteenth century, the *Koch* tribes had established a dynasty in the Bihar region, but were driven from power by the Muslims and the Ahoms. Today, they prefer the title *Rajbansi*, which means 'men of royal blood', because of their ruling ancestors.
112.	*Kodagu*	The *Kodagu* (also known as the *Coorg*) live in the Kodagu (Coorg) district of Karnataka. The word kodagu means 'situated to the West', and their district, Coorg, stands in the ridges of the Western Ghat Mountains. They speak a Dravidian language that is also called *Kodagu*.
113.	*Kadamukudy*	These tribes are found in Kadamukudy Island, near Kerala.
114.	*Kodavas*	The *Kodavas* are the people living in the Kodagu region of Karnataka, which lies in the Western Ghats.
115.	*Kokni*	Also known as *Kukna*, *Kanara* and *Kokna*; these tribes are found in the Western Ghats of Maharashtra and in the Dang district of Gujarat. Their language is known as *Kukna*.
116.	*Kolam*	Known by the names such as *Kolamboli*, *Kulme* and *Kolmi*. These tribes mainly inhabit the hills and plains of Andhra Pradesh and Maharashtra. They also live in the Yavatmal, Wardha and Nanded districts of Maharashtra and parts of Madhya Pradesh. They speak a Dravidian language called *Kolami*, but can also speak *Marathi*, *Telugu* or *Gondi*. Around the twelfth century, the *Kolam* served as priests for the *Gonds*, representing some of their important Gods.
117.	*Koli*	These tribes are found in Dadra and Nagar Haveli.
118.	*Konyaks*	These tribes are found in the Mon district of Nagaland. *Konyaks* are adept artisans and skilled craftsmen. These tribes are ruled by hereditary chiefs known as Anghs. The most colourful festival of the *Konyaks* is called Aoling Monyu and it is observed during the first week of April every year.

S.N.	Name of Tribe	Distribution and Major Characteristics
119.	*Koras*	These tribes are found in Bihar.
120.	*Korku*	Known variously as *Bondeya, Bopchi, Korki, Kurku* and *Kuri,* these tribes are found in the Southern Betul, Hoshangabad and East Nimar districts of Southern Madhya Pradesh, and Amravati, Buldhana and Akola districts of Northern Maharashtra. Their language is also known as *Korku.*
121.	*Korwa*	The *Korwas* are one of the STs of Central India. They live in the hills, valleys, and forests of Uttar Pradesh, Bihar and Madhya Pradesh. They speak a *Munda* language, also called *Korwa,* which belongs to the Austro–Asiatic language family. There are two very distinct tribes among the *Korwas*: the *Diharia* (or *Kisan*), who are farmers, and the *Paharia* (or *Benwaria*), who live in the hills. These two tribes do not inter-marry.
122.	*Kota*	*Kota* or *Kotta* are the tribes found in the Nilgiri Hills of Kotagiri, in Tamil Nadu. Their language is also known as *Kota.*
123.	*Kotwali*	These tribes are found in Gujarat.
124.	*Koya*	The *Koyas* live in the forests, plains, and valleys on both sides of the Godavari river in Andhra Pradesh. Many also live in Madhya Pradesh and Odisha. The language of these tribes is also called *Koya* and is closely related to *Gondi* and has been strongly influenced by *Telugu.*
125.	*Kui*	Known by various names such as *Kuy, Kodi, Khond, Khondi, Khondo, Kanda, Kodulu* and *Kuinga,* these tribes are found in the Udayagiri area of Odisha, and parts of Andhra Pradesh, Madhya Pradesh and Tamil Nadu. They are a sub-group of a Dravidian tribe known as the *Kond,* which literally means 'mountaineers'. They speak *Kui,* a Dravidian language.
126.	*Kuki*	These tribes are found in the Kohima district of Nagaland and parts of Manipur and Assam.
127.	*Kulu Pahari*	Known by names such as *Kauli, Kulu Boli, Kulu Pahari, Kulu Pahari, Kullui Pahar, Pahari Kullu, Phari Kulu, Kulvi, Kullui* and *Kulwali,* these tribes are found primarily in Himachal Pradesh. They speak *Kului* and *Hindi.*
128.	*Kumaoni*	These tribes are also known as *Kumauni, Kumau, Kumawani, Kumgoni, Kumman* and *Kunayaoni* and are found in Assam, Bihar, Delhi, Madhya Pradesh, Uttar Pradesh, Maharashtra and Nagaland.
129.	*Kurichiya*	These tribes are found in the forest areas near Calicut, Tellichery and Waynad in Kerala where they were the earliest known inhabitants. They are also found in parts of Tamil Nadu. Their language is called *Kurichiya,* although many also speak Malayalam. It is said that their name comes from the word *kuri,* which is the sandalwood that they put on their foreheads and chests.
130.	*Kurumba*	These tribes are mainly concentrated in the Dharmapuri district of Tamil Nadu and also found in parts of Maharashtra, Karnataka, Andhra Pradesh and Kerala. They are also known as *Kuruba, Kurumban, Kurumvari* and Southern *Kanares.* The *Kuruba* are considered closely related to the

S.N.	Name of Tribe	Distribution and Major Characteristics
		Pallavas of the eighth century. As the *Pallava* rule declined, the *Kuruba's* forefathers scattered over a wide area of Southern India and became culturally distinct. They speak a Dravidian language that is called *Kurumba* or Southern Kannada.
131.	*Labans*	Described variously as *Banjaras*, *Lambadi*, *Gola*, *Wanji* and *Singali*, these tribes are found in Andhra Pradesh, Madhya Pradesh, Himachal Pradesh, Gujarat, Tamil Nadu, Maharashtra, Karnataka, Odisha and West Bengal. They speak a language called *Lamani*.
132.	*Labbai*	These people are found in Tamil Nadu and claim to be the descendants of Arab traders.
133.	*Ladakhi*	These tribes are found in Ladakh in Jammu and Kashmir.
134.	*Lahauli*	These tribes are found in Lahaul Valley of Himachal Pradesh.
135.	*Lalung*	The *Lalung* tribe lives in the States of Assam and Meghalaya. They are divided into two main sub-divisions: those who live in the hills and those who live on the plains. The two have developed distinctions due to differences in terrain and ecology. The *Lalung* belong to a race known as the *Bodo*. According to legend, their ancestors migrated from Tibet and followed the path of the Brahmaputra river south into the hills of Assam. Later, perhaps during the mid-1600's they migrated to the plains.
136.	*Lambada*	These tribes are known by various names such as *Lambadi*, *Lavani*, *Lemadi*, *Lumadale*, *Banjara*, *Bangala*, *Banjori*, *Banjuri*, *Gohar– Herkeri*, *Goola*, *Gurmarti*, *Kora*, *Gormati*, *Singali*, *Sugalis*, *Tanda*, *Vanjara* and *Wanji*. The *Lambadas* originated in Rajasthan, but are now concentrated in Andhra Pradesh. These are also found in Madhya Pradesh, Himachal Pradesh, Gujarat, Tamil Nadu, Maharashtra, Karnataka, Odisha and West Bengal. They speak *Lamani* and *Telugu*.
137.	*Lepcha*	Known by different names such as *Rong*, *Rongke*, *Rongpa* and *Nunpa*, the people of this tribe are found in Sikkim, Kalimpong district of West Bengal, Nepal and Bhutan. Their language is known as *Lepcha*.
138.	*Lhoba*	These tribes, which are found mostly in Arunachal Pradesh, are part of the *Lhoba* tribe of China. Their language, *Lhoba*, is part of the Sino–Tibetan language family.
139.	*Lohar*	These tribes are found in Rajasthan, Gujarat, Madhya Pradesh, Maharashtra, Uttar Pradesh, Delhi, Haryana and Punjab. They are known by various names such as *Gade Lohar*, *Gaduliya Lohar*, *Lohpitta Rajput Lohar*, *Bagri Lohar* and *Bhubaliya Lohar*. Their language is known as *Gade Lohar*.
140.	*Lothas*	*Lothas* are the tribes found in the Wokha district of Nagaland. They are known for their colourful dances and songs particularly Shanta, Tokhu Emong and Pikhuchak.
141.	*Lushi*	The *Lushi* (*Shenlung*) tribes are found in Mizoram and claim Jewish ancestry.

S.N.	Name of Tribe	Distribution and Major Characteristics
142.	*Madiga*	These people belong to Andhra Pradesh and are known by different names such as *Gentoo*, *Tailangi*, *Telangire*, *Telgi*, *Tengu*, *Terangi* and *Tolangan*.
143.	*Malakkaras*	The word 'Malakkara' means 'mighty people of the forests'. These tribes, which are also known as *Mala Muthas*, are found in Kerala. They speak *Malamutha* and Malayalam.
144.	*Malappanackers*	These are the tribes found in Kerala. They speak *Malappanackan* and Malayalam.
145	*Malmis*	These tribes are found in Lakshadweep.
146.	*Malto*	Known as *Malti*, *Maltu*, *Malpaharia* and *Rajmahalia*, these tribes are found in Rajmahal hills in North-east Bihar and West Bengal. Their language is also called *Malto*.
147.	*Manipuri Jews*	This community sees itself as descendants of the *Menashe* Tribe (which is considered as one of the 10 lost tribes of Jews). These people have Chinese appearance and they claim that after their forefathers were exiled and enslaved by the Assyrians they somehow escaped from slavery and moved initially to China and later to India. Known as the 'Children of Menmasseh', they live on the border of India and Burma, and a majority of them are believed to be Messianic Jews (believers in Jesus Christ - Yeshua). Their number is estimated to be about 1.2-1.5 million. Israelis refer to these people by the name *Shinlung* or *Lushi*.
148.	*Manjhi*	These tribes live mainly in the Gumla district of Bihar and parts of West Bengal and Assam. Their primary language is *Manjhi*, but they also speak *Sadri* and Bihari.
149.	*Mappilla (Moplas)*	These are the Muslim tribes found in Kerala's northern coastal towns and the Lakshadweep Islands off the Malabar Coast. The Laccadive *Mappilla* are unique, in that they have maintained a matrilineal society (line of descent is traced through the women) in spite of their Islamic beliefs.
150.	*Maria*	The *Maria* or *Muria* are a sub-group of *Gonds* and live in the forests and hills of Central India. They live along the Indrawati river, which flows through the Bastar district of Madhya Pradesh. They are divided into two groups: the *Maria* and the *Bison Horn Maria*. The two groups have very little contact with each other, except during certain annual festivals.
151.	*Mauris*	These tribes are found in Jammu and Kashmir.
152.	*Mavchi*	The *Mavchis* are a sub-group of *Bhils*. Their language is also called *Mavchi* or *Mawchi*.
153.	*Mech*	The *Mech* are mainly found in the forests and hills of Darjeeling, Jalpaiguri, and Coochbehar districts of West Bengal and also parts of Assam. They are related to the *Bodo* and *Kachari* tribes and speak *Bodo*, a branch of the Sino–Tibetan language family.
154.	*Meithei*	These tribes are also known as *Mitei*, *Manipuris*, *Kathe*, *Kathi* and *Ponna*, and are found in Assam, Manipur, Nagaland, Tripura, Uttar Pradesh and West Bengal, besides Bangladesh and Myanmar. These tribes are of Mongol origin and speak a Tibeto–Burman language called *Meithei*.

S.N.	Name of Tribe	Distribution and Major Characteristics
155.	*Melacheris*	These tribes are found in Lakshadweep.
156.	*Meos*	These are the rural tribes concentrated in the districts of Alwar and Bharatpur in Rajasthan and in the Gurgaon district in Haryana. They speak in *Marwari* (*Rajasthani*).
157.	*Mikir*	These tribes, which are also known by the names *Mikiri*, *Manchati* and *Karbi*, are found in the Mikir and Rengma Hills of Assam. Their language is called *Mikir*.
158.	*Mina*	These tribes are found in Rajasthan and Madhya Pradesh.
159.	*Minicoy*	These tribes are found in the Minicoy islands of Lakshadweep.
160.	*Mishing*	These tribes are found in the Majuli Island of Arunachal Pradesh. They are also known as *Miri*. They speak *Miri*, a dialect of *Adi*.
161.	*Mishmi*	These tribes are found in Arunachal Pradesh and Assam.
162.	*Mog*	These are the Buddhist tribes found in Tripura, who claim to have originated from Arakan.
163.	*Mon*	These tribes are found in the Ladakh region.
164.	*Monpa*	Known by different names such as *Moinba*, *Monba*, *Mompa*, *Momba*, *Menpa* and *Memba*; these tribes are found in Arunachal Pradesh and parts of Tibet. They speak a language called *Moinba*.
165.	*Mru*	The *Mrus* are primarily located in the region where the borders of India, Bangladesh and Myanmar intersect. They are concentrated in the Chittagong Hills in Bangladesh, Arakan Yoma district of Myanmar and in the Jalpaiguri district of West Bengal. The native language of these tribes is also called *Mru*.
166.	*Mundas*	These tribes are also known by different names such as *Mundari*, *Mandari*, *Munari*, *Horo*, *Mondari* and *Colh*. These are known as *Adivasis* in Odisha. They are found mainly in Southern and Western parts of the Ranchi district of Bihar and also in the States of Assam, Himachal Pradesh, Madhya Pradesh, Odisha, Tripura and West Bengal, and in Andaman and Nicobar Islands, besides Nepal and Bangladesh. Their language is called Mundari. Mundas are the most ancient among the tribes of Bihar.
167.	*Nahari*	They are located mainly in the Raipur, Bilaspur, and Sambalpur districts of Madhya Pradesh, as well as parts of Odisha.
168.	*Naikda*	These tribes are found in Dadra and Nagar Haveli, and Daman and Diu.
169.	*Neethakani*	These tribes are found mainly in the Neelwai village of Andhra Pradesh.
170.	*Nicobarese*	These are the people of Mongoloid stock found on the Car Nicobar Island. These tribes are also known as *Car* and *Pu*.
171	*Nihal*	These tribes are found in Maharashtra and Madhya Pradesh, and are also known as *Nihali*, *Nahal*, *Nahali* and *Kalto*.
172.	*Nimadi*	These tribes are located mainly in the States of Madhya Pradesh, Uttar Pradesh and Maharashtra.

S.N.	Name of Tribe	Distribution and Major Characteristics
173.	*Nishi*	These tribes are found in Arunachal Pradesh.
174.	*Nocte*	These are the tribes found in Nagaland and are known by other names such as *Borduria, Jaipuria, Paniduria, Mohongia* and *Namsangia.*
175.	*Noniya Caste*	The *Noniyas* are the traditional earth-workers of Uttar Pradesh.
176.	*Ojhas*	These tribes are found in Madhya Pradesh.
177.	*Onges*	These tribes are found on the Southern Andaman Islands, Little Andaman Island and the three small islands to the North-east. They are also known as *Jarawas* and speak the *Jarawa* language.
178.	*Oraon (Kurukh)*	These tribes are found in Bihar and Odisha.
179.	*Oriya*	These tribes are found mainly in Odisha and parts of Bihar, West Bengal, Assam and Andhra Pradesh. They are also known as *Odri, Odrum, Ouya, Odisha, Uriya, Utkali, Vadiy* and *Yudhia.*
180.	*Padhar*	These tribes are found in Gujarat.
181.	*Paharis*	These tribes are found along the foothills of the Himalayas, mainly in Himachal Pradesh.
182.	*Pangwalis*	These tribes are found in Himachal Pradesh.
183.	*Panika*	These tribes live primarily in the Shahdol district of Madhya Pradesh and are divided into two broad groups: the *Kabirpanthi* and the *Sakta.*
184.	*Paniyan*	These tribes are found in Kerala and Tamil Nadu and are also known as *Pania* and *Panyah.*
185.	*Parhiya*	These are mainly found on the hilltop forests of Bihar.
186.	*Pardhi*	These tribes are known by different names as *Bahelia, Chita Pardhi, Lango Pardhi, Paidia, Paradi, Paria, Phans Pardhi, Takankar* and *Takia.* These tribes are found in Andhra Pradesh, Madhya Pradesh, Gujarat and Maharashtra. Their language is also described as Pardhi.
187.	*Pauri Bhuiya*	These tribes, who are also known *Bhuiya, Bhuinhar, Bhuinya, Bhuiyali, Bhumia, Bhungiyas* and *Bhuyan Oriya,* are located in Bihar, Madhya Pradesh, Maharashtra, Odisha, Uttar Pradesh and West Bengal.
188.	*Pawi*	These tribes are found in Mizoram.
189.	*Phoms*	These tribes are found in the Tuensang district of Nagaland.
190.	*Poouch*	These tribes are found in Jammu and Kashmir. They speak *Poochi, Urdu* and *Dogri* languages.
191.	*Pramalai Kallar*	These tribes are found in Tamil Nadu.
192.	*Porja*	The *Porja* (or *Konda–Dora*) tribes are located in the States of Andhra Pradesh, Assam, and Odisha. They speak a Dravidian language called *Konda–Dora,* (*konda* meaning 'hill' and *dora* meaning 'leader'). This name was given to them because of their descent from the great *Kondh* tribe. The *Porja* are well established as two groups: the *Pedda Kondalu* and *Chinna Kondalu.* The *Pedda Kondalu* have retained most aspects of their traditional culture, however, the *Chinna Kondalu* group has been heavily influenced by the *Telugu.*

S.N.	Name of Tribe	Distribution and Major Characteristics
193.	*Purig–Pa*	These are people of Tibetan origin found in the Kargil district of Jammu and Kashmir. The people prefer to be culturally and linguistically identified with Tibet, although religiously with Islam.
194.	*Rabari*	These are semi-nomadic people of Rajasthan.
195.	*Rabha*	These tribes are found in Assam.
196.	*Raji*	These tribes are found in Uttar Pradesh. Their language is also known as *Raji*.
197.	*Ralte*	These tribes are found in Mizoram.
198.	*Reddy*	These people are found mainly in Andhra Pradesh and Tamil Nadu. They are also known by other names such as *Mukha–Rora, Deddi–Dora* and *Conta–Reddi*.
199.	*Rengma*	These tribes are found in Tseminyu in the Kohima district of Nagaland. The *Rengmas* celebrate the popular festival of Ngada every year in November.
200.	*Riang*	The *Riangs* are primarily located in the Deotamura ranges of eastern and central Tripura. Some tribes also live in Assam. These are considered as the second largest tribe of Tripura and the first inhabitants there. They speak a Baric language that is also called *Riang*.
201.	*Saharia*	The *Saharia* or *Sor* live primarily in the State of Madhya Pradesh. They are also found in the hills of the Ganjam district of Southern Odisha, as well as in the States of Andhra Pradesh, Bihar, Tamil Nadu, West Bengal and Assam. They speak a *Munda* language that belongs to the Austro–Asiatic language family.
202.	*Sangtams*	These tribes are found in the Tuensang district of Nagaland.
203.	*Santhals*	These tribes are found in Bihar, Odisha, Tripura, West Bengal and Assam, besides Bangladesh and Nepal. They are also known by other names such as *Hor*, *Har*, *Satar*, *Santhali*, *Sandal*, *Sangtal*, *Sentali*, *Samtali* and *Santhial*. They speak *Santhali*. They are considered as the largest tribe in India. The basis of *Santhal* religion is the belief that they are totally surrounded by bongas or spirits and frequently are visited by deceased ancestors.
204.	*Saurashtras*	*Saurashtrians* are mostly silk weavers and silk thread merchants, originated in the Saurashtra region (present day Gujarat, and parts of Maharashtra) in Northern India and later settled in Madurai and spread to Trichy, Salem, Kumbakonam and other surrounding areas of Tamil Nadu few centuries ago. Their native language, also known as *Sourashtra*, is a pre-*Gujarati* language, which shows many archaic features of *Gujarati* and at the same time shows influence of other Aryan languages such as *Konkani* and *Marathi* and of Dravidian *Telugu* and *Tamil*.
205.	*Sema*	*Semas* are considered to be the martial tribe among the *Nagas*. These tribes are concentrated in the Zunheboto district of Nagaland. Tuluni is one of the most important festivals observed by the *Sema* tribe in the second week of July every year.

S.N.	Name of Tribe	Distribution and Major Characteristics
206.	*Sentenelese*	These are the tribes found in the Andaman and Nicobar Islands.
207.	*Sherdukpen*	These tribes are found in Arunachal Pradesh.
208.	*Sherpa*	The *Sherpa* are Himalayan people, living primarily in eastern Nepal. A few of them, however, live in the mountains of the bordering Indian State of Sikkim and in the larger hill towns of the State of West Bengal. The *Sherpa* language is a dialect of Tibetan, thus it is a part of the Tibeto–Burman language family.
209.	*Shin*	These people are found in the Dras Valley and Gurais area in Kishenganga Valley in Northern Kashmir. They speak in an Indo–Aryan language called *Shina*.
210.	*Shompens (Shom Peng)*	These tribes live on the Great Nicobar Island. Their language is also described as *Shom Peng*.
211.	*Siddis*	These tribes are found in Goa.
212.	*Singpho*	These tribes are found in Arunachal Pradesh.
213.	*Solung*	These tribes are found in Arunachal Pradesh.
214.	*Sondwari*	These are the tribes found mainly in Madhya Pradesh and Rajasthan.
215.	*Sourastras*	Described by various names such as *Saurashtra*, *Saurashtri* and *Patnuli*; these tribes are found in parts of Tamil Nadu. Their spoken language is described as *Saurashtra*.
216.	*Subernabaniks*	They represent the rich merchant classes of West Bengal. The *Subernabanik* women are famous for their beauty. According to legends, the *Subernabaniks* originated in the princely kingdom of Oudh, near Bengal, during the reign of King Adisura about 1000 AD.
217.	*Tadvi Bhil*	They live in Gujarat, Maharashtra, Karnataka and Rajasthan. Their language, known as *Dhanka*, belongs to the *Bhil* sub-group of the Indo–Aryan linguistic family. The name 'Tadvi' literally means 'one who taps the palm tree' and is used to refer to all forest tribes.
218.	*Tagin*	This tribe is found in Assam. They speak a language, which is also called *Tagin*. Some also speak *Dafla* or *Nisi* as a trade language.
219.	*Tangsa*	These are the tribes found in Nagaland and in Myanmar. They are known by various names such as *Rangpan*, *Tasey* and *Chem Chang*. Their language is *Naga* and *Tangsa*.
220.	*Tharu*	These tribes are found in the North-west of Bihar and speak *Tharu*, a dialect of Bhojpuri.
221.	*Toda*	These are the hill tribes of Nilgiris (Tamil Nadu) who are very indigenous to this area. These people, who are also known as *Todi* or *Tuda*, claim to be the direct descendants of the Aryans. According to some researchers, this tribe has links with the Sumerian people of the ancient Mesopotamian civilization. Their language is also known as *Toda*. Today, almost 3,000 *Todas* live in the Blue Mountains.

S.N.	Name of Tribe	Distribution and Major Characteristics
222.	*Totos*	These tribes are found in West Bengal.
223.	*Tripuris (Tipera)*	These tribes are primarily concentrated in the plains of Tripura and are indigenous to this area. They are the largest tribe in the State, and their land once extended over much of Bengal, Assam and Northern Myanmar.
224.	*Tulu*	These are fair-skinned people who are found mainly in Mangalore and South Kannara districts of Karnataka. They speak a Dravidian language, also called *Tulu*, which lacks a script.
225.	*Ullada*	These tribes are found in Kerala.
226.	*Urali*	These tribes are found in Kerala and Tamil Nadu. These are also known by the names *Oorazhi* and *Urli,* and speak the *Urali* language.
227.	*Usipi*	The *Usipis* are located predominantly in Assam and Tripura. Their language, *Usipi* (also known as *Kok Barok*), is part of the Sino–Tibetan language family. The *Usipis* are commonly characterized as Garo.
228.	*Warli*	These tribes are found in Dadra and Nagar Haveli, and Daman and Diu.
229.	*Wagdi*	The *Wagdis* are considered as a sub-group of *Bhils*. The *Wagdi* language, also called *Wagdi*, belongs to the Bhil branch of the Indo–Aryan language family.
230.	*Wancho*	Known by different names such as *Naga Wancho*, *Banpara Naga* and *Joboka*, these tribes are found in Assam, Arunachal Pradesh and Nagaland. They speak *Naga* and *Wancho* languages.
231.	*Wars*	These tribes primarily inhabit the districts east of Meghalaya. They belong to the Proto–Austroloid Mon–Khmer race.
232.	*Wattal*	These tribes are found in Jammu and Kashmir.
233.	*Yanadi*	The *Yanadi* are dark-skinned short-stature people found mainly in the Nellore and Chittoor districts of Andhra Pradesh. They speak *Yanadi*, which belongs to the *Telugu* language family. They are divided into two groups: the *Manchi Yanadi*, which belongs to the small superior class and the *Challa Yanadi*. The two groups seldom inter-marry.
234.	*Yeravas*	These tribes are found in Karnataka.
235.	*Yerukala*	The *Yerukala* live primarily in the Godivari district of Andhra Pradesh and parts of Tamil Nadu and Kerala. They speak *Sankara–Yerukala*, a Southern Dravidian language that is distinctly related to *Tamil*. Many of the *Yerukala* also speak *Telugu*. These tribes are also known by the names *Yarukula*, *Yerukla*, *Erukala* and *Koraya*.
236.	*Yimchungers*	These tribes are found in the Tuensang District of Nagaland.
237.	*Zakhring*	These tribes are found in Arunachal Pradesh.
238.	*Zanskaris*	These are the Buddhist tribes, ethnically similar to Tibetans, found in the Zanskar Mountains in the Leh–Ladakh and Kargil–Purik areas of Jammu and Kashmir. They speak *Zanskari*, a language related to Tibetan.
239.	*Zemi Naga*	The *Zemi Nagas* or *Zeliang* are the tribes found in the Upper Barak Valley of Assam and parts of Manipur and Nagaland. They speak *Naga* and *Zeme* (*Zoliang*) languages.

Source: Culturopedia, 2014

ANNEXURE II

Table 21: List of notified Indian tribes

Sr. No.	State/Union Territory and Notified Indian Tribes
	Andhra Pradesh
1.	*Andh, Sadhu Andh*
2.	*Bagata*
3.	*Bhil*
4.	*Chenchu*
5.	*Gadabas, Bodo Gadaba, Gutob Gadaba, Kallayi Gadaba, Parangi Gadaba, Kathera Gadaba, Kapu Gadaba*
6.	*Gond, Naikpod, Rajgond, Koitur*
7.	*Goudu* (in the Agency tracts)
8.	*Hill Reddis*
9.	*Jatapus*
10.	*Kammara*
11.	*Kattunayakan*
12.	*Kolam, Kolawar*
13.	*Konda Dhoras, Kubi*
14.	*Konda Kapus*
15.	*Kondareddis*
16.	*Kondhs, Kodi, Kodhu, Desaya Kondhs, Dongria Kondhs, Kuttiya Kondhs, Tikiria Kondhs, Yenity, Kondhs, Kuvinga*
17.	*Kotia, Bentho Oriya, Bartika, Holva, Sanrona, Sidhopaiko*
18.	*Koya, Doli Koya, Gutta Koya, Kammara Koya, Musara Koya,Oddi Koya, Pattidi Koya, Rajah, Rasha Koya, Lingadhari Koya (ordinary), Kottu Koya, Bhine Koya, Rajkoya.*
19.	*Kulia*

Sr. No.	State/Union Territory and Notified Indian Tribes
20.	*Malis* (excluding Adilabad, Hyderabad, Karimnagar, Khammam, Mahbubnagar, Medak, Nalgonda, Nizamabad and Warangal districts)
21.	*Manna Dhora*
22.	*Mukha Dhora, Nooka Dhora*
23.	*Nayaks* (in the Agency tracts)
24.	*Pardhan*
25.	*Porja, Parangiperja*
26.	*Reddi Dhoras*
27.	*Rona, Rena*
28.	*Savaras, Kapu Savaras, Maliya Savaras, Khutto Savaras*
29.	*Sugalis, Lambadis, Banjara*
30.	*Thoti* (in Adilabad, Hyderabad, Karimnagar, Khammam, Mahbubnagar, Medak Nalgonda, Nizamabad and Warangal districts)
31.	*Valmiki* (Scheduled Areas of Vishakhapatnam, Srikakulam, Vijayanagram, East Godavari and West Godavari districts)
32.	*Yenadis, Chella Yenadi, Kappala Yenadi, Manchi Yenadi, Reddi Yenadi*
33.	*Yerukulas, Koracha, Dabba Yerukula, Kunchapuri Yerukula, Uppu Yerukula*
34.	*Nakkala, Kurvikaran*
35.	*Dhulia, Paiko, Putiya* (in the districts of Vishakhapatnam and Vijayanagaram)
	Arunanchal Pradesh
1.	*Abor*
2.	*Aka*
3.	*Apatani*
4.	*Nyishi*
5.	*Galo*
6.	*Khampti*
7.	*Khowa*
8.	*Mishmi, Idu, Taroan*
9.	*Momba*
10.	*Any Naga tribes*
11.	*Sherdukpen*
12.	*Singpho*
13.	*Hrusso*
14.	*Tagin*
15.	*Khamba*

Sr. No.	State/Union Territory and Notified Indian Tribes
16.	*Adi*
	Assam
	In the autonomous districts of Karbi Anglong and North Cachar Hills.
1.	*Chakma*
2.	*Dimasa, Kachari*
3.	*Garo*
4.	*Hajong*
5.	*Hmar*
6.	*Khasi, Jaintia, Synteng, Pnar, War, Bhoi, Lyngngam*
7.	Any *Kuki* tribes, including (i) *Biate, Biete* (ii) *Changsan* (iii) *Chongloi* (iv) *Doungel* (v) *Gamalhou* (vi) *Gangte* (vii) *Guite* (viii) *Hanneng* (ix) *Haokip Haupit* (x) *Haolai* (xi) *Hengna* (xii) *Hongsungn* (xiii) *Hrangkhwal, Rangkhol* (xiv) *Jongbe* (xv) *Khawchung* (xvi) *Khawathlang, Khothalong* (xvii) *Khelma* (xviii) *Kholhou* (xix) *Kipgen* (xx) *Kuki* (xxi) *Lengthang* (xxii) *Lhangum* (xxiii) *Lhoujem* (xxiv) *Lhouvun* (xxv) *Lupheng* (xxvi) *Mangjel* (xxvii) *Misao* (xxviii) *Riang* (xxix) *Sairhem* (xxx) *Sairhem* (xxxi) *Selnam* (xxxii) *Singson* (xxxiii) *Sitlhou* (xxxiv) *Sukte* (xxxv) *Thado* (xxxvi) *Thangngeu* (xxxvii) *Uibuh* (xxxviii) *Vaiphei*
8.	*Lakher*
9.	Man (Tai speaking)
10.	Any *Mizo* (*Lushai*) tribes
11.	Karbi
12.	Any *Naga* tribes
13.	*Pawi*
14.	*Syntheng*
15.	*Lalung*
	In the State of Assam including the districts in the Bodo land territorial areas and excluding the autonomous districts of Karbi Anglong and North Cachar Hills
16.	*Barmans* in Cachar
17.	*Boro, Borokachar*
18.	*Deori*
19.	*Hojai*
20.	*Kachari*
21.	*Sonwal*
22.	*Mech*
23.	*Miri*
24.	*Rabha*

Sr. No.	State/Union Territory and Notified Indian Tribes
25.	*Dimasa*
26.	*Hajong*
27.	*Singhpho*
28.	*Khampti*
29.	*Garo*
30.	*Bihar*
31.	*Asur, Agaria*
32.	*Baiga*
33.	*Banjara*
34.	*Bathudi*
35.	*Bedia*
36.	*Binjhia*
37.	*Birhor*
38.	*Birjia*
39.	*Chero*
40.	*Chik Baraik*
41.	*Gond*
42.	*Gorait*
43.	*Ho*
44.	*Karmali*
45.	*Kharia,Dhelki Kharia Dudh Kharia, Hill Kharia*
46.	*Kharwar*
47.	*Khond*
48.	*Kisan, Nagesia*
49.	*Kora, Mudi-Kora*
50.	*Korwa*
51.	*Lohara, Lohra*
52.	*Mahli*
53.	*Mal Paharia, Kumarbhag Paharia*
54.	*Munda, Patar*
55.	*Oraon, Dhangar (Oraon)*
56.	*Parhaiya*
57.	*Santal*
58.	*Sauria Paharia*

Sr. No.	State/Union Territory and Notified Indian Tribes
59.	*Savar*
60.	*Kawar*
61.	*Kol*
62.	*Tharu*
	Chhattisgarh
1.	*Agariya*
2.	*Andh*
3.	*Baiga*
4.	*Bhaina*
5.	*Bharia Bhumia, Bhuinhar Bhumia, Bhumiya, Bharia, Paliha, Pando*
6.	*Bhattra*
7.	*Bhil, Bhilala, Barela, Patelia*
8.	*Bhil Mina*
9.	*Bhunjia*
10.	*Biar, Biyar*
11.	*Binjhwar*
12.	*Birhul, Birhor*
13.	*Damor, Damaria*
14.	*Dhanwar*
15.	*Gadaba, Gadba*
16.	*Gond, Arakh, Arrakh, Agaria, Asur, Badi Maria*
17.	*Halba, Halbi*
18.	*Kamar*
19.	*Karku*
20.	*Kawar, Kanwar, Kaur, Cherwa, Rathia, Tanwar, Chhatri*
21.	*Khairwar, Kondar*
22.	*Kharia*
23.	*Kondh, Khond,Kandh*
24.	*Kol*
25.	*Kolam*
26.	*Korku, Bopchi, Mouasi, Nihal, Nahul Bondhi, Bondeya*
27.	*Korwa, Kodaku*
28.	*Majhi*
29.	*Majhwar*

Sr. No.	State/Union Territory and Notified Indian Tribes
30.	Mawasi
31.	*Munda*
32.	*Nagesia, Nagasia*
33.	*Oraon, Dhanka, Dhangad*
34.	*Pao*
35.	*Pardhan, Pathari, Saroti*
36.	*Pardhi, Bahelia, Bahellia, Chita Pardhi, Langoli Pardhi, Phans Pardhi, Shikari, Takankar, Takia* [In (i) Bastar, Dantewara, Kanker, Raigarh, Jashpurnagar, Surguja and Koria districts, and (ii) Katghora, Pali, Kartala and Korba tahsils of Korba district, (iii) Bilaspur, Pendra, Kota and Takhatpur tahsils of Bilaspur district, (iv) Durg, Patan Gunderdehi, Dhamdha, Balod, Gurur and Dondilohara tahsils of Durg district, (v) Chowki, Manpur and Mohala Revenue Inspector Circles of Rajnandgaon district, (vi) Mahasamund Saraipali and Basna tahsils of Mahasamund district,(vii) Bindra-Navagarh Rajim and Deobhog tahsils of Raipur district, (viii) Dhamtari, Kurud and Sihava tahsils of Dhamtari district.
37.	*Parja*
38.	*Sahariya, Saharia, Seharia, Sehria, Sosia, Sor*
39.	*Saonta, Saunta*
40.	*Saur*
41.	*Sawar, Sawara*
42.	*Sonr*
	Goa
1.	*Dhodia*
2.	*Dubla (Halpati)*
3.	*Naikda (Talavia*
4.	*Siddi (Nayaka)*
5.	*Warli*
6.	*Kunbi*
7.	*Gawda*
8.	*Velip*
	Gujarat
1.	*Barda*
2.	*Bavacha, Bamcha*
3.	*Bharwad* (in the Nesses of the forests of Alech, Barada and Gir)
4.	*Bhil, Bhil Garasia, Dholi Bhil, Dungri Bhil, Dungri Garasia, Mewasi Bhil, Rawal Bhil, Tadvi Bhil, Bhagalia, Bhilala, Pawra, Vasava, Vasave*
5.	*Charan* (in the Nesses of the forests of Alech, Barada and Gir)

Sr. No.	State/Union Territory and Notified Indian Tribes
6.	*Chaudhri* (in Surat and Valsad districts)
7.	*Chodhara*
8.	*Dhanka,Tadvi, Tetaria, Valvi*
9.	*Dhodia, Dhodi*
10.	*Dubla, Talavia, Halpati*
11.	*Gamit, Gamta, Gavit, Mavchi, Padvi*
12.	*Gond, Rajgond*
13.	*Kathodi, Katkari, Dhor Kathodi, Dhor Katkari, Son Kathodi, Son Katkari*
14.	*Kokna, Kokni, Kukna*
15.	*Koli Dhor, Tokre Koli, Kolcha, Kolgha*
16.	*Kunbi* (in the Dangs district)
17.	*Naikda, Nayaka, Cholivala Nayaka, Kapadia Nayaka, Mota Nayaka, Nana Nayaka*
18.	*Padhar*
19.	*Pardhi*
20.	*Advichincher, Phanse Pardhi* (excluding Amreli, Bhavanagar, Jamnagar, Junagadh, Kutch, Rajkot and Surendranagar districts)
21.	*Patelia, Pomla*
22.	*Rabari* (in the Nesses of the forests of Alech, Barada and Gir)
23.	*Rathawa*
24.	*Siddi, Siddi-Badshan* (in Amreli, Bhavnagar, Jamnagar, Junagadh, Rajkot and Surendranagar districts)
25.	*Warli*
26.	*Vitola, Kotwalia, Barodia*
27.	*Bhil, Bhilala, Barela, Patelia*
28.	*Tadvi Bhil, Bawra,Vasave*
29.	*Padvi*
	Himachal Pradesh
1.	*Bhot, Bodh*
2.	*Gaddi*
3.	*Gujjar*
4.	*Jad, Lamba, Khampa*
5.	*Kanaura, Kinnara*
6.	*Lahaula*
7.	*Pangwala*
8.	*Swangla*

Sr. No.	State/Union Territory and Notified Indian Tribes
9.	*Beta, Beda*
10.	*Domba, Gara, Zoba*
	Jammu and Kashmir
1.	*Balti*
2.	*Beda*
3.	*Bot, Boto*
4.	*Brokpa, Drokpa, Dard, Shin*
5.	*Changpa*
6.	*Garra*
7.	*Mon*
8.	*Purigpa*
9.	*Gujjar*
10.	*Bakarwal*
11.	*Gaddi*
12.	*Sippi*
	Jharkhand
1.	*Asur*
2.	*Agaria*
3.	*Baiga*
4.	*Banjara*
5.	*Bathudi, Bedia*
6.	*Binjhia*
7.	*Birhor*
8.	*Birjia*
9.	*Chero*
10.	*Chik Baraik*
11.	*Gond*
12.	*Gorait*
13.	*Ho*
14.	*Karmali*
15.	*Kharia, Dhelki Kharia, Dudh Kharia, Hill Kharia*
16.	*Kharwar*
17.	*Khond*
18.	*Kisan, Nagesia*

Sr. No.	State/Union Territory and Notified Indian Tribes
19.	*Kora, Mudi-Kora*
20.	*Korwa*
21.	*Lohra*
22.	*Mahli*
23.	*Mal Paharia, Kumarbhag Paharia*
24.	*Munda, Patar*
25.	*Oraon, Dhangar(Oraon)*
26.	*Parhaiya*
27.	*Santal*
28.	*Sauria Paharia*
29.	*Savar*
30.	*Bhumij*
31.	*Kawar*
32.	*Kol*
	Karnataka
1.	*Adiyan*
2.	*Barda*
3.	*Bavacha, Bamcha*
4.	*Bhil, Bhil Garasia, Dholi Bhil, Dungri Bhil, Dungri Garasia, Mewasi Bhil, Rawal Bhil, Tadvi Bhil, Bhagalia, Bhilala, Pawra, Vasava, Vasave*
5.	*Chenchu, Chenchwar*
6.	*Chodhara*
7.	*Dubla, Talavia, Halpati*
8.	*Gamit, Gamta, Gavit, Mavchi, Padvi, Valvi*
9.	*Gond, Naikpod, Rajgond*
10.	*Gowdalu*
11.	*Hakkipikki*
12.	*Hasalaru*
13.	*Irular*
14.	*Iruliga*
15.	*Jenu Kuruba*
16.	*Kadu Kuruba*
17.	*Kammara* (in South Kanara district and Kollegal taluka of Mysore district)
18.	*Kaniyan, Kanyan* (in Kollegal taluka of Mysore district)
19.	*Kathodi, Katkari, Dhor Kathodi, Dhor Katkari, Son Kathodi, Son Katkari*

Sr. No.	State/Union Territory and Notified Indian Tribes
20.	*Kattunayakan*
21.	*Kokna, Kokni, Kukna*
22.	*Koli Dhor, Tokre Koli, Kolcha, Kolgha*
23.	*Konda Kapus*
24.	*Koraga*
25.	*Kota*
26.	*Koya, Bhine Koya, Rajkoya*
27.	*Kudiya, Melakudi*
28.	*Kuruba* (in Coorg district)
29.	*Kurumans*
30.	*Maha Malasar*
31.	*Malaikudi*
32.	*Malasar*
33.	*Malayekandi*
34.	*Maleru*
35.	*Maratha* (in Coorg district)
36.	*Marati* (in South Kanara district)
37.	*Meda, Medara, Medari, Gauriga, Burud*
38.	*Naikda, Nayaka, Chollivala Nayaka, Kapadia Nayaka, Mota Nayaka, Nana Nayaka, Naik, Nayak, Beda, Bedar*, and *Valmiki*
39.	*Palliyan*
40.	*Paniyan*
41.	*Pardhi, Advichincher, Phanse Pardhi, Haranshikari*
42.	*Patelia*
43.	*Rathawa*
44.	*Sholaga*
45.	*Soligaru*
46.	*Toda*
47.	*Warli*
48.	*Vitolia, Kotwalia, Barodia*
49.	*Yerava*
50.	*Siddi* (in Uttar Kannada district)
	Kerala
1.	*Adiyan*
2.	*Arandan, Aranadan3*

Sr. No.	State/Union Territory and Notified Indian Tribes
3.	*Eravallan*
4.	*Hill Pulaya, Mala Pulayan, Kurumba Pulayan, Karavazhi Pulayan, Pamba Pulayan*
5.	*Irular, Irulan*
6.	*Kadar, Wayanad Kadar*
7.	*Kanikaran, Kanikkar*
8.	*Kattunayakan*
9.	*Kochuvelan*
10.	*Koraga*
11.	*Kudiya, Melakudi*
12.	*Kurichchan, Kurichiyan*
13.	*Kurumans, Mullu Kuruman, Mulla Kuruman, Mala Kuruman*
14.	*Kurumbas, Kurumbar, Kurumban*
15.	*Maha Malasar*
16.	*Malai Arayan, Mala Arayan*
17.	*Malai Pandaram*
18.	*Malai Vedan, Malavedan*
19.	*Malakkuravan*
20.	*Malasar*
21.	*Malayan, Nattu Malayan, Konga Malayan* (excluding the areas comprising the Kasargode, Connanore, Wayanad and Kozhikode districts)
22.	*Malayarayar*
23.	*Mannan*
24.	*Muthuvan, Mudugar, Muduvan*
25.	*Palleyan, Palliyan, Palliyar, Paliyan*
26.	*Paniyan*
27.	*Ulladan, Ullatan*
28.	*Uraly*
29.	*Mala Vettuvan* (in Kasargode and Kannur districts)
30.	*Ten Kurumban, Jenu Kurumban*
31.	*Thachanadan, Thachanadan Moopan*
32.	*Cholanaickan*
33.	*Mavilan*
34.	*Karimpalan*
35.	*Vetta Kuruman*
36.	*Mala Panickar*

Sr. No.	State/Union Territory and Notified Indian Tribes
	Madhya Pradesh
1.	*Agariya*
2.	*Andh*
3.	*Baiga*
4.	*Bhaina*
5.	*Bharia Bhumia, Bhuinhar Bhumia, Bhumiya, Bharia, Paliha, Pando*
6.	*Bhattra*
7.	*Bhil, Bhilala, Barela, Patelia*
8.	*Bhil Mina*
9.	*Bhunjia*
10.	*Biar, Biyar*
11.	*Binjhwar*
12.	*Birhul, Birhor*
13.	*Damor, Damaria*
14.	*Dhanwar*
15.	*Gadaba, Gadba*
16.	*Gond; Arakh, Arrakh, Agaria, Asur, Badi Maria, Bada Maria, Bhatola, Bhimma, Bhuta, Koilabhuta, Koilabhuti, Bhar, Bisonhorn Maria, Chota Maria, Dandami Maria, Dhuru, Dhurwa, Dhoba, Dhulia, Dorla, Gaiki, Gatta, Gatti, Gaita, Gond Gowari, Hill Maria, Kandra, Kalanga, Khatola, Koitar, Koya, Khirwar, Khirwara, Kucha Maria, Kuchaki Maria, Madia, Maria, Mana, Mannewar, Moghya, Mogia, Monghya, Mudia, Muria, Nagarchi, Nagwanshi, Ojha, Raj, Sonjhari Jhareka, Thatia, Thotya, Wade Maria, Vade Maria, Daroi*
17.	*Halba, Halbi*
18.	*Kamar*
19.	*Karku*
20.	*Kawar, Kanwar, Kaur*
21.	*Cherwa, Rathia, Tanwar, Chattri*
22.	*Khairwar, Kondar*
23.	*Kharia*
24.	*Kondh, Khond, Kandh*
25.	*Kol*
26.	*Kolam*
27.	*Korku, Bopchi, Mouasi, Nihal, Nahul Bondhi, Bondeya*
28.	*Korwa, Kodaku*
29.	*Majhi*

Sr. No.	State/Union Territory and Notified Indian Tribes
30.	*Majhwar*
31.	*Mawasi*
32.	*Munda*
33.	*Nagesia, Nagasia*
34.	*Oraon, Dhanka, Dhangad*
35.	*Panika* [in (i) Chhatarpur, Panna, Rewa, Satna, Shahdol, Umaria, Sidhi and Tikamgarh districts, and (ii) Sevda and Datia tehsils of Datia district]
36.	*Pao*
37.	*Pardhan, Pathari, Saroti*
38.	*Pardhi, Bahelia, Bahellia, Chita Pardhi, Langoli Pardhi, Phans Pardhi, Shikari, Takankar, Takia* [In (i) Chhindwara, Mandla, Dindori and Seoni districts, (ii) Baihar Tahsil of Balaghat District, (iii) Betul, Bhainsdehi and Shahpur tahsils of Betul district, (iv) Patan tahsil and Sihora and Majholi blocks of Jabalpur district, (v) Katni (Murwara) and Vijaya Raghogarh tahsils and Bahoriband and Dhemerkheda blocks of Katni district, (vi) Hoshang abad, Babai, Sohagpur, Pipariya and Bankhedi tahsils and Kesla block of Hoshangabad district, (vii) Narsinghpur district, and (viii) Harsud Tahsil of Khandwa district]
39.	*Parja*
40.	*Sahariya, Saharia, Seharia, Sehria, Sosia, Sor*
41.	*Saonta, Saunta*
42.	*Saur*
43.	*Sawar*
44.	*Sawara*
45.	*Sonr*
	Maharashtra
1.	*Andh*
2.	*Baiga*
3.	*Barda*
4.	*Bavacha, Bamcha*
5.	*Bhaina*
6.	*Bharia Bhumia, Bhuinhar Bhumia, Pando*
7.	*Bhattra*
8.	*Bhil, Bhil Garasia, Dholi Bhil, Dungri Bhil, Dungri Garasia, Mewasi Bhil, Rawal Bhil, Tadvi Bhil, Bhagalia, Bhilala Pawra, Vasava, Vasave*
9.	*Bhunjia*
10.	*Binjhwar*
11.	*Birhul, Birhor*

Sr. No.	State/Union Territory and Notified Indian Tribes
12.	*Dhanka, Tadvi, Tetaria, Valvi*
13.	*Dhanwar*
14.	*Dhodia*
15.	*Dubla, Talavia, Halpati*
16.	*Gamit, Gamta, Gavit, Mavchi, Padvi*
17.	*Gond, Rajgond, Arakh, Arrakh, Agaria, Asur, Badi Maria, Bada Maria, Bhatola, Bhimma, Bhuta, Koilabhuta, Koilabhuti, Bhar, Bisonhorn Maria, Chota Maria, Dandami Maria, Dhuru, Dhurwa, Dhoba, Dhulia, Dorla, Gaiki, Gatta, Gatti, Gaita, Gond Gowari, Hill Maria, Kandra, Kalanga, Khatola, Koitar, Koya, Khirwar, Khirwara, Kucha Maria, Kuchaki Maria, Madia, Maria, Mana, Man newar, Moghya, Mogia, Monghya, Mudia, Muria, Nagarchi, Naikpod, Nag wanshi, Ojha, Raj, Sonjhari Jhareka, Thatia, Thotya, Wade Maria, Vade Maria*
18.	*Halba, Halbi*
19.	*Kamar*
20.	*Kathodi, Katkari, Dhor Kathodi, Dhor Kathkari*
21.	*Son Kathodi, Son Katkari*
22.	*Kawar, Kanwar, Kaur, Cherwa, Rathia, Tanwar, Chattri*
23.	*Khairwar*
24.	*Kharia*
25.	*Kokna, Kokni, Kukna*
26.	*Kol*
27.	*Kolam, Mannervarlu*
28.	*Koli Dhor; Tokre Koli, Kolcha, Kolgha*
29.	*Koli Mahadev, Dongar Koli*
30.	*Koli Malhar*
31.	*Kondh, Khond, Kandh*
32.	*Korku, Bopchi, Mouasi, Nihal, Nahul, Bondhi, Bondeya*
33.	*Koya, Bhine Koya, Rajkoya*
34.	*Nagesia, Nagasia*
35.	*Naikda, Nayaka, Cholivala Nayaka, Kapadia Nayaka, Mota Nayaka, Nana Nayaka*
36.	*Oraon, Dhangad*
37.	*Pardhan, Pathari, Saroti*
38.	*Pardhi, Advichincher, Phans Pardhi, Phanse Pardhi, Langoli Pardhi, Bahelia, Bahellia, Chita Pardhi, Shikari, Takankar, Takia*
39.	*Parja*
40.	*Patelia*

Sr. No.	State/Union Territory and Notified Indian Tribes
41.	*Pomla*
42.	*Rathawa*
43.	*Sawar, Sawara*
44.	*Thakur, Thakar, Ka Thakur, Ka Thakar, Ma Thakur, Ma Thakar*
45.	*Warli*
46.	*Vitolia, Kotwalia, Barodia*
	Manipur
1.	*Aimol*
2.	*Anal*
3.	*Angami*
4.	*Chiru*
5.	*Chothe*
6.	*Gangte*
7.	*Hmar*
8.	*Kabui, Inpui, Rongmei*
9.	*Kacha Naga, Liangmai, Zeme*
10.	*Koirao,Thangal*
11.	*Koireng*
12.	*Kom*
13.	*Lamgang*
14.	*Mao*
15.	*Maram*
16.	*Maring*
17.	Any *Mizo* (*Lushai*) Tribes
18.	*Monsang*
19.	*Moyon*
20.	*Paite*
21.	*Purum*
22.	*Ralte*
23.	*Sema*
24.	*Simte*
25.	*Suhte*
26.	*Tangkhul*
27.	*Thàdou*

Sr. No.	State/Union Territory and Notified Indian Tribes
28.	*Vaiphui*
29.	*Zou*
30.	*Poumai Naga*
31.	*Tarao*
32.	*Kharam*
33.	Any *Kuki* tribes
34.	*Mate*
	Meghalaya
1.	*Chakma*
2.	*Dimasa, Kachari*
3.	*Garo*
4.	*Hajong*
5.	*Hmar*
6.	*Khasi, Jaintia, Synteng, Pnar, War, Bhoi, Lyngngam*
7.	Any *Kuki* tribes, including i. *Biate, Biete* ii. *Changsan* iii. *Chongloi* iv. *Doungel* v. *Gamalhou* vi. *Gangte* vii. *Guite* viii. *Hanneng* ix. *Haokip, Haupit* x. *Haolai* xi. *Hengna* xii. *Hongsungh* xiii. *Hrangkhwal, Rangkhol* xiv. *Jongbe* xv. *Khawchung* xvi. *Khawathlang, Khothalong* xvii. *Khelma* xviii. *Kholhou* xix. *Kipgen* xx. *Kuki* xxi. *Lengthang* xxii. *Lhangum* xxiii. *Lhoujem* xxiv. *Lhouvun* xxv. *Lupheng* xxvi. *Mangjel* xxvii. *Misao* xxviii. *Riang* xxix. *Sairhem* xxx. *Selnam* xxxi. *Singson* xxxii. *Sitlhou* xxxiii. *Sukte* xxxiv. *Thado* xxxv. *Thangngeu* xxxvi. *Uibuh* xxxvii. *Vaiphei.*
8.	*Lakher*
9.	*Man* (Tai Speaking)
10.	Any *Mizo* (*Lushai*) tribes
11.	*Mikir*
12.	Any *Naga* tribes
13.	*Pawi*
14.	*Synteng*
15.	*Boro Kacharis*
16.	*Koch*
17.	*Raba, Rava*
	Mizoram
1.	*Chakma*
2.	*Dimasa (Kachari)*
3.	*Garo*
4.	*Hajong*

Sr. No.	State/Union Territory and Notified Indian Tribes
5.	*Hmar*
6.	*Khasi* and *Jaintia*, (Including *Khasi, Synteng* or *Pnar, War, Bhoi* or *Lyngngam*)
7.	Any *Kuki* tribes, including (i) *Baite or Beite* (ii) *Changsan* (iii) *Chongloi* (iv) *Doungel* (v) *Gamalhou* (vi) *Gangte* (vii) *Guite* (viii) *Hanneng* (ix) *Haokip* or *Haupit* (x) *Haolai* (xi) *Hengna* (xii) *Hongsungh* (xiii) *Hrangkhwal* or *Rangkhol* (xiv) *Jongbe* (xv) *Khawchung* (xvi) *Khawathlang* or *Khothalong* (xvii) *Khelma* (xviii) *Kholhou* (xix) *Kipgen* (xx) *Kuki* (xxi) *Lengthang* (xxii) *Lhangum* (xxiii) *Lhoujem* (xxiv) *Lhouvun* (xxv) *Lupheng* (xxvi) *Mangjel* (xxvii) *Missao* (xxviii) *Riang* (xxix) *Siarhem* (xxx) *Selnam* (xxxi) *Singson* (xxxii) *Sitlhou* (xxxiii) *Sukte* (xxxiv) *Thado* (xxxv) *Thangngeu* (xxxvi) *Uibuh* (xxxvii) *Vaiphei*
8.	Lakher
9.	*Man* (Tai-speaking)
10.	Any *Mizo* (*Lushai*) tribes
11.	*Mikir*
12.	Any *Naga* tribes
13.	*Pawi*
14.	*Synteng*
15.	*Paite*
	Nagaland
1.	*Naga*
2.	*Kuki*
3.	*Kachari*
4.	*Mikir*
5.	*Garo*
	Odisha
1.	*Bagata, Bhakta*
2.	*Baiga*
3.	*Banjara, Banjari*
4.	*Bathudi, Bathuri*
5.	*Bhottada, Dhotada Bhotra, Bhatra, Bhattara, Bhotora, Bhatara*
6.	*Bhuiya, Bhuyan*
7.	*Bhumia*
8.	*Bhumij, Teli Bhumij, Haladipokhria Bhumij, Haladi Pokharia Bhumija, Desi Bhumij, Desia Bhumij, Tamaria Bhumij*
9.	*Bhunjia*
10.	*Binjhal, Binjhwar*

Sr. No.	State/Union Territory and Notified Indian Tribes
11.	*Binjhia, Binjhoa*
12.	*Birhor*
13.	*Bondo Poraja, Bonda Paroja, Banda Paroja*
14.	*Chenchu*
15.	*Dal*
16.	*Desua Bhumij*
17.	*Dharua, Dhuruba, Dhurva*
18.	*Didayi, Didai Paroja, Didai*
19.	*Gadaba, Bodo Gadaba, Gutob Gadaba, Kapu Gadaba, Ollara Gadaba, Parenga Gadaba, Sano Gadaba*
20.	*Gandia*
21.	*Ghara*
22.	*Gond, Gondo Rajgond, Maria Gond, Dhur Gond*
23.	*Ho*
24.	*Holva*
25.	*Jatapu*
26.	*Juang*
27.	*Kandha Gauda*
28.	*Kawar Kanwar*
29.	*Kharia, Kharian Berga Kharia, Dhelki Kharia, Dudh Kharia, Erenga Kharia, Munda Kharia,Oraon Kharia, Khadia, Pahari Kharia*
30.	*Kharwar*
31.	*Khond, Kond, Kandha, Nanguli Kandha, Sitha Kandha Kondh, Kui, Buda Kondh, Bura Kandha, Desia Kandha, Dungaria Kondh, Kutia Kandha, Kandha Gauda, Muli Kondh, Malua Kondh, Pengo Kandha, Raja Kondh, Raj Khond*
32.	*Kisan, Nagesar, Nagesia*
33.	*Kol*
34.	*Kolah Laharas, Kol Loharas*
35.	*Kolha*
36.	*Koli, Malhar*
37.	*Kondadora*
38.	*Kora, Khaira, Khayara*
39.	*Korua*
40.	*Kotia*
41.	*Koya, Gumba Koya, Koitur Koya, Kamar Koya, Musara Koya*

Sr. No.	State/Union Territory and Notified Indian Tribes
42.	*Kulis*
43.	*Lodha, Nodh, Nodha, Lodh*
44.	*Madia*
45.	*Mahali*
46.	*Mankidi*
47.	*Mankirdia, Mankria, Mankidi*
48.	*Matya, Matia*
49.	*Mirdhas, Kuda, Koda*
50.	*Munda, Munda Lohara, Munda Mahalis, Nagabanshi Munda, Oriya Munda*
51.	*Mundari*
52.	*Omanatya, Omanatyo, Amanatya*
53.	*Oraon, Dhangar, Uran*
54.	*Parenga*
55.	*Paroja, Parja, Bodo Paroja, Barong Jhodia Paroja, Chhelia Paroja, Jhodia Paroja, Konda Paroja, Paraja, Ponga Paroja, Sodia Paroja, Sano Paroja, Solia Paroja*
56.	*Pentia*
57.	*Rajuar*
58.	*Santal*
59.	*Saora, Savar, Saura, Sahara Arsi Saora, Based Saora, Bhima Saora, Bhimma Saora, Chumura Saora, Jara Savar, Jadu Saora, Jati Saora, Juari Saora, Kampu Saora, Kampa Soura, Kapo Saora, Kindal Saora, Kumbi Kancher Saora, Kalapithia Saora, Kirat Saora, Lanjia Saora, Lamba Lanjia Saora, Luara Saora, Luar Saora, Laria Savar, Malia Saora, Malla Saora, Uriya Saora, Raika Saora, Sudda Saora, Sarda Saora, Tankala Saora, Patro Saora, Vesu Saora.*
60.	*Shabar*
61.	*Lodha*
62.	*Sounti*
63.	*Tharua, Tharua Bindhani*
	Rajasthan
1.	*Bhil, Bhil Garasia, Dholi Bhil, Dungri Bhil, Dungri Garasia, Mewasi Bhil, Rawal Bhil, Tadvi Bhil, Bhagalia, Bhilala, Pawra, Vasava, Vasave*
2.	*Bhil Mina*
3.	*Damor, Damaria*
4.	*Dhanka, Tadvi, Tetaria, Valvi*
5.	*Garasia* (excluding *Rajput Garasia*)
6.	*Kathodi, Katkari, Dhor Kathodi, Dhor Katkari, Son Kathodi, Son Katkari*

Sr. No.	State/Union Territory and Notified Indian Tribes
7.	*Kokna, Kokni, Kukna*
8.	*Koli Dhor, Tokre Koli, Kolcha, Kolgha*
9.	*Mina*
10.	*Naikda, Nayaka, Cholivala Nayaka, Kapadia Nayaka, Mota Nayaka, Nana Nayaka*
11.	*Patelia*
12.	*Seharia, Sehria, Sahariya*
	Sikkim
1.	*Bhutia* (including *Chumbipa, Dopthap, Dukpa, Kagatey, Sherpa, Tibetan, Tromop, Yolmo*)
	Lepcha Limboo Tamang
	Tamil Nadu
1.	*Adiyan*
2.	*Aranadan*
3.	*Eravallan*
4.	*Irular*
5.	*Kadar*
6.	*Kammara* (excluding Kanyakumari district and Shenkottah taluk of Tirunelveli district)
7.	*Kanikaran, Kanikkar* (in Kanyakumari district and Shenkottah and Ambasam udram taluks of Tirunelveli District)
8.	*Kaniyan, Kanyan*
9.	*Kattunayakan*
10.	*Kochu Velan*
11.	*Konda Kapus*
12.	*Kondareddis*
13.	*Koraga*
14.	*Kota* (excluding Kanyakumari district and Shenkottah taluk of Tirunelveli district)
15.	*Kudiya, Melakudi*
16.	*Kurichchan*
17.	*Kurumbas* (in the Nilgiris district)
18.	*Kurumans*
19.	*Maha Malasar*
20.	*Malai Arayan*
21.	*Malai Pandaram*
22.	*Malai Vedan*

Sr. No.	State/Union Territory and Notified Indian Tribes
23.	*Malakkuravan*
24.	*Malasar*
25.	*Malayali* (in Dharmapuri, North Arcot Pudukottai, Salem, South Arcot and Tiruchirapali districts)
26.	*Malayekandi*
27.	*Mannan*
28.	*Mudugar, Muduvan*
29.	*Muthuvan*
30.	*Palleyan*
31.	*Palliyan*
32.	*Palliyar*
33.	*Paniyan*
34.	*Sholaga*
35.	*Toda* (excluding Kanyakumari district and Shenkottah Taluk of Tirunelveli district)
36.	*Uraly*
	Tripura
1.	*Bhil*
2.	*Bhutia*
3.	*Chaimal*
4.	*Chakma*
5.	*Garoo*
6.	*Halam, Bengshel, Dub, Kaipeng, Kalai, Karbong, Lengui, Mussum, Rupini, Sukuchep, Thangchep*
7.	*Jamatia*
8.	*Khasia*
9.	*Kuki*, including the following sub-tribes: (i) *Balte* (ii) *Belalhut* (iii) *Chhalya* (iv) *Fun* (v) *Hajango* (vi) *Jangtei* (vii) *Khareng* (viii) *Khephong* (ix) *Kuntei* (x) *Laifang* (xi) *Lentei* (xii) *Mizel* (xiii) *Namte* (xiv) *Paitu, Paite* (xv) *Rangchan* (xvi) *Rangkhole* (xvii) *Thangluya*
10.	*Lepcha*
11.	*Lushai*
12.	*Mag*
13.	*Munda, Kaur*
14.	*Noatia, Murashing*
15.	*Orang*
16.	*Riang*

Sr. No.	State/Union Territory and Notified Indian Tribes
17.	*Santal*
18.	*Tripura, Tripuri, Tippera*
19.	*Uchai*
	Uttarakhand
1.	*Bhotia*
2.	*Buksa*
3.	*Jaunsari*
4.	*Raji*
5.	*Tharu*
	Uttar Pradesh
1.	*Bhotia*
2.	*Buksa*
3.	*Jaunsari*
4.	*Raji*
5.	*Tharu*
6.	*Gond, Dhuria, Nayak, Ojha, Pathari, Raj Gond* (in the districts of Mehrajganj, Sidharth Nagar, Basti, Gorakhpur, Deoria, Mau, Azamgarh, Jonpur, Balia, Gazipur, Vara nasi, Mirzapur and Sonbhadra)
7.	*Kharwar, Khairwar* (in the districts of Deoria, Balia, Ghazipur, Varanasi and Sonbhadra)
8.	*Saharya* (in the district of Lalitpur)
9.	*Parahiya* (in the district of Sonbhadra)
10.	*Baiga* (in the district of Sonbhadra)
11.	*Pankha, Panika* (in the districts of Sonbhadra and Mirzapur)
12.	*Agariya* (in the district of Sonbhadra)
13.	*Patari* (in the district of Sonbhadra)
14.	*Chero* (in the districts of Sonbhadra and Varanasi)
15.	*Bhuiya, Bhuinya* (in the district of Sonbhadra)
	West Bengal
1.	*Asur*
2.	*Baiga*
3.	*Bedia, Bediya*
4.	*Bhumij*
5.	*Bhutia,Sherpa, Toto, Dukpa, Kagatay, Tibetan, Yolmo*
6.	*Birhor*

Sr. No.	State/Union Territory and Notified Indian Tribes
7.	*Birjia*
8.	*Chakma*
9.	*Chero*
10.	*Chik Baraik*
11.	*Garo*
12.	*Gond*
13.	Gorait
14.	*Hajang*
15.	*Ho*
16.	*Karmali*
17.	*Kharwar*
18.	*Khond*
19.	*Kisan*
20.	*Kora*
21.	*Korwa*
22.	*Lepcha*
23.	*Lodha, Kheria, Kharia*
24.	*Lohara, Lohra*
25.	*Magh*
26.	*Mahali*
27.	*Mahli*
28.	*Mal Pahariya*
29.	*Mech*
30.	*Mru*
31.	*Munda*
32.	*Nagesia*
33.	*Oraon*
34.	*Parhaiya*
35.	*Rabha*
36.	*Santal*
37.	*Sauria Paharia*
38.	*Savar*
39.	*Limbu (Subba)*
40.	*Tamang*

Sr. No.	State/Union Territory and Notified Indian Tribes
	Andaman and Nicobar Islands
1.	*Andamanese, Chariar, Chari, Kora, Tabo, Bo, Yere, Kede, Bea, Balawa, Bojigiyab, Juwai, Kol*
2.	*Jarawas*
3.	*Nicobarese*
4.	*Onges*
5.	*Sentinelese*
6.	*Shom Pens*
	Dadra and Nagar Haveli
1.	*Dhodia*
2.	*Dubla* including *Halpati*
3.	*Kathodi*
4.	*Kokna*
5.	*Koli Dhor* including *Kolgha*
6.	*Naikda* or *Nayaka*
7.	*Warli*
	Daman and Diu
1.	*Dhodia*
2.	*Dubla (Halpati)*
3.	*Naikda (Talavia)*
4.	*Siddi (Nayaka)*
5.	*Warli*
	Lakshadweep
1.	Inhabitants of the Laccadive, Minicoy and Aminidivi Islands who, and both of whose parents, were born in those islands. 'Provided that the children who are born to inhabitants of Lakshadweep in any other place in the mainland of India shall be deemed to be inhabitants born in the islands if such children settle permanently in the islands'. Explanation: The term 'settle permanently' shall have the same meaning as defined under Clause 3 (I) (d) of the Lakshadweep Panchayat Regulation, 1994.

Source: Annual Report - 2012-13, MoTA, Government of India.

ANNEXURE III

Table 22: Indian tribes and their common dialects

S.N.	Languages/Dialect	Tribes and Distribution
1.	Aariya	Madhya Pradesh.
2.	Abujmaria *(Hill Maria)*	This is a Dravidian language spoken in Abujmar Hills in Bastar District of Madhya Pradesh by the *Hill Maria* tribes.
3.	Adi *(Abor, Abor–Miri)*	It is a Tibeto–Burman language spoken by people in northern hills of the Assam valley, between Bhutan and the Buruli river and Arunachal Pradesh. *Miri* and *Mishing* are its dialects.
4.	Adiyan *(Eravas)*	It is a Dravidian language spoken mainly in the Kannur District of Kerala and parts of Tamil Nadu and Karnataka.
5.	Agariya *(Agoria)*	This is spoken by tribals in Bihar, Madhya Pradesh, Maharashtra and Odisha.
6.	Ahom *(Tai Ahom)*	This language is no longer spoken in daily life, but is used in religious chants and literary materials by tribes in Assam.
7.	Aimol	This language is spoken by some tribes in Assam and Manipur. *Langrong* is its dialect.
8.	Aiton	Spoken in Assam
9.	Ajmeri	It is a dialect of Rajasthani, spoken mainly in Ajmer.
10.	Aka–Bea *(Bea, Bojigniji, Bojigyab)*	This language is spoken by tribes of the coastal areas of South Andaman Island.
11.	Aka–Bo *(Bo, Ba)*	Spoken in east-central coast of North Andaman Island and the North Reef Island
12.	Aka–Cari *(Chariar)*	Spoken in the north coast of North Andaman Island, Landfall Island and other nearby small islands
13.	Aka–Jeru *(Jeru, Yerawa)*	Spoken in parts of the interior and South North Andaman Island
14.	Aka–Kede *(Kede)*	Spoken in parts of the Middle Andaman Island

S.N.	Languages/Dialect	Tribes and Distribution
15.	Aka–Kol *(Kol)*	Spoken in south-east Middle Andaman Island
16.	Aka–Kora *(Kora)*	Spoken in north-east and north-central coasts of North Andaman Island and Smith Island
17.	Akar–Bale *(Bale, Balwa)*	Spoken in Andaman Islands, Ritchie's Archipelago, Havelock Island and Neil Island
18.	Allar *(Chatans)*	It is a Dravidian language spoken mainly in the Palghat region of Tamil Nadu.
19.	Alu *(Alu Kurumba)*	It is Dravidian language spoken in the Nilgiris district of Tamil Nadu.
20.	Amwi	This language is spoken by some tribes in the Khasi Hills.
21.	Anal *(Namfau)*	Spoken in the South-east Manipur and parts of Myanmar and Bangladesh
22.	Andh *(Andhi)*	Spoken in parts of Andhra Pradesh, Madhya Pradesh and Maharashtra
23.	Anga *(Angika)*	Spoken in parts of Bihar
24.	Apatani *(Apa)*	Spoken in parts of Assam, Arunachal Pradesh and Nagaland
25.	A–Pucikwar *(Puchikwar)*	Spoken in parts of Andaman Islands, Boratang Island, South coast of Middle Andaman Island, north-east coast of South Andaman Island
26.	Arakanese *(Mogh, Magh, Maghi, Morma, Yakan, Rakhain, Marma)*	Spoken in Bangladesh, Myanmar and parts of Assam, Tripura and Mizoram
27.	Arakh *(Arrakh)*	Spoken in parts of Madhya Pradesh and Maharashtra
28.	Aranadan *(Eranadans)*	Spoken in parts of Tamil Nadu and Karnataka and in the Calicut region of Kerala
29.	Are *(Ade Bhasha, Aray, Arya)*	Spoken in parts of Andhra Pradesh, Maharashtra and Karnataka
30.	Asuri *(Ashree, Asura)*	It is spoken in parts of Bihar, Madhya Pradesh, Maharashtra, Odisha and West Bengal. *Brijia* (*Koranti*) and *Manjhi* are its dialects.
31.	Awadhi *(Abadi, Abohi, Ambodhi, Baiswari, Kosali)*	It is a dialect of *Hindi* spoken in parts of Bihar, Madhya Pradesh, Uttar Pradesh, Delhi and parts of Nepal.
32.	Badaga (*Badugu*)	It is a Dravidian language spoken in the Nilgiri hills and Kunda hills of Tamil Nadu.
33.	Bagata (*Bhakta*)	Spoken in parts of Andhra Pradesh, Tamil Nadu and Odisha
34.	Bagheli (*Bagelkhandi, Mannadi, Riwai, Ganggai, Mandal,Kewat, Nagpuri Marathi*)	It is spoken in the North-east Madhya Pradesh and parts of Maharashtra, Uttar Pradesh and Nepal.

S.N.	Languages/Dialect	Tribes and Distribution
35.	Bagri (*Bagria, Baorias*)	Spoken in parts of Punjab, Rajasthan, Punjab, Haryana and Madhya Pradesh by the *Bagri* tribesmen. It is spoken mainly by the *Bansphor, Nat, Bishnoi, Jat* and *Lohar* tribes
36.	Bahawalpuri (Riasati)	Spoken in some parts of Madhya Pradesh and Maharashtra
37.	Baiga (*Bega, Bhumia*)	Spoken in some parts of Bihar, Maharashtra, Madhya Pradesh, Odisha and West Bengal by the *Baiga* tribes
38.	Bangaru	This is an Indo–Aryan language spoken by over 5.7 million Bangris located mainly in the states of Haryana, Karnataka, Himachal Pradesh, Uttar Pradesh, Punjab and Delhi.
39.	Bareli	Bareli or Barel is spoken by the people of the *Barel* tribes in some parts of Madhya Pradesh and in Northern Dhule district of Maharashtra. The language belongs to the *Bhil* branch of the Indo–Aryan language family and reflects both *Rajasthani* and *Gujarati* influence.
40.	Bateri	Spoken in some places near Srinagar in Jammu and Kashmir
41.	Bathudi	Spoken in some parts of Bihar, Odisha and West Bengal by the *Bathudi* tribes
42.	Bauria (*Badak, Babri, Basria, Vaghri, Baori*)	The Bauria language is spoken by the Bauria tribes and belongs to the *Bhil* branch of the Indo–Aryan language family and reflects both *Rajasthani* and *Gujarati* influence. It is spoken in some parts of Gujarat, Himachal Pradesh, Madhya Pradesh, Rajasthan and Uttar Pradesh.
43.	Bazigar	Spoken in some parts of Gujarat, Himachal Pradesh, Jammu and Kashmir, Madhya Pradesh and Karnataka
44.	Bedia (*Bediya, Beriya*)	Spoken in some parts of Bihar, Gujarat, Madhya Pradesh, Uttar Pradesh, Maharashtra, Karnataka and West Bengal
45.	Bellari	It is a Dravidian language related to *Tulu* and *Koraga*
46.	Bhadrawahi (*Baderwali, Badrohi, Bahi*)	Spoken in some parts of Jammu and Kashmir; *Bhalesi* and *Padari* are its dialects while *Pangi* is a language similar to it.
47.	Bhalay	It is spoken in many parts of the Amravati district in Maharashtra.
48.	Bharia (*Bhar, Bharat, Bhumia, Bhumiya, Paliha*)	It is spoken in Bilaspur, Chhatarpur, Chhindwara, Datia, Durg, Jabalpur, Mandla, Panna, Rewa, Sidhi, Surguja, Tikamgarh districts of Madhya Pradesh and also in some parts of Uttar Pradesh and West Bengal.
49.	Bhattiyali (*Bhatiali Pahari, Bhatiyali*)	It is spoken in Himachal Pradesh, mainly in parts of the Chamba district.
50.	Bhatneri	It is spoken mainly in parts of north-west Bhattiana region of Madhya Pradesh.

S.N.	Languages/Dialect	Tribes and Distribution
51.	Bhatola	Spoken in some parts of Madhya Pradesh
52.	Bhatri (*Bhattri, Bhattra,Bhatra*)	Spoken in some parts of Andhra Pradesh, Madhya Pradesh, Maharashtra and Odisha
53.	Bhilala (*Bhilali*)	Spoken in some parts of Gujarat, Madhya Pradesh, Maharashtra, Karnataka and Rajasthan
54.	Bhili (*Bhilbari, Bhilboli, Bhilla, Bhil, Vil*)	Spoken in some parts of Andhra Pradesh, Madhya Pradesh, Bihar, Gujarat, Jammu and Kashmir, Maharashtra, Karnataka, Punjab, Rajasthan and Tripura. It has various dialects such as Anarya, Baoria, Bhimchaura, Charani, Dhodhri, Haburi, Kokna, Kotali, Kotvali (Kotwalia), Magri, Naiki, Neelishikari, Padvi, Panchali, Pittala Bhasha, Ranawat, Rani, Siyalgiri, Tadavi, Takankari, Takari and Valvi.
55.	Bhilori	It is spoken in parts of Northern Dhule district, around Dhadgaon in Maharashtra. *Bhilodi* and *Noiri* (Satpuda Noiri) are its dialects.
56.	Bhim (*Bhina*)	Spoken in some parts of Tripura
57.	Bhojpuri (*Bhojapuri, Bihari*)	Spoken in parts of the Purnea district in Bihar and some parts of Assam, Delhi, Madhya Pradesh, Uttar Pradesh and West Bengal. It has several dialects such as Northern Standard Bhojpuri (Gorakhpuri, Sarawaria, Basti), Western Standard Bhojpuri (Purbi, Benarsi), Southern Standard Bhojpuri (Kharwari), Tharu, Madhesi, Domra and Musahari.
58.	Bhottara	Spoken in some parts of Andhra Pradesh, Tamil Nadu and Odisha
59.	Bhoyari (*Bhomiyari, Bhuria*)	Spoken in some parts of Maharashtra. Kumbhari, Marari, Ojhi, Powari are its dialects.
60.	Bhuiya (*Bhuinhar, Bhuinya, Bhumia*)	Spoken in some parts of Bihar, Madhya Pradesh, Maharashtra, Odisha, Uttar Pradesh and West Bengal
61.	Bhumij (*Sadar Bhumij, Munda*)	Spoken primarily in Mayurbhanj District, Odisha and Singhbhum district, Bihar and also in parts of West Bengal. Its dialects are *Kisan–Bhumij*, *Kurmi*, *Parse–Khumij* and *Rahiya*.
62.	Bhunjia (*Bhumjiya*)	Spoken in Raipur, Hoshangabad, Sambalpur, Kalshandi districts of Madhya Pradesh and in parts of Maharashtra and Odisha
63.	Biete (*Bete*)	Spoken in some parts of Meghalaya, Mizoram, Assam and Cachar Hills
64.	Bihari	The heartland for Bihari lies in the Bihar Province of India and extends over into Western Terai of Southern Nepal. The Bihari is spoken in several dialects, the most common of

S.N.	Languages/Dialect	Tribes and Distribution
		which is Bhojpuri. Although this dialect has very little written literature, it has many oral folk tales. Other Bihari dialects are Nagpuri and Magahi.
65.	Bijori (*Binjhia, Birijia, Burja)*	Spoken in some parts of Bihar, Madhya Pradesh, Odisha and West Bengal
66.	Bilaspuri (*Pacchmi, Kahluri, Kehluri*)	Spoken in some parts of Himachal Pradesh
67.	Binjhwari (*Binjhal*)	This language is spoken by the Binjhwari or Binjhal tribes in some parts of Bihar, Madhya Pradesh, Maharashtra and West Bengal.
68.	Birhor (*Bihor, Birhar, Birhore*)	Spoken in parts of Hazaribagh and Ranchi districts and the Singbhum area of Bihar and parts of Madhya Pradesh, Maharashtra, Odisha and West Bengal
69.	Bodo (*Boro, Bodi, Bara, Boroni, Mechi*)	Spoken in parts of Assam, West Bengal and Nepal, Chote and Mech are its dialects.
70.	Bodo Parja (*Parji, Parja, Jodia Parja*)	Spoken in some parts of Koraput district in Odisha
71.	Bondo (*Bhonda, Gutobi, Remo, Nanqa Poroja)*	Spoken in some parts of Koraput and Bondo Hills in Odisha
72.	Braj Bhasha (*Braj, Antarbedi*)	It is a dialect of *Hindi* and is spoken in some parts of Bihar, Madhya Pradesh, the Agra region in Uttar Pradesh and parts of Rajasthan and Delhi. *Bhuksa*, *Sikarwari*, *Jadobafi* and *Dangi* are its dialects.
73.	Brokskat (*Brokpa, Brokpa of Dah–Hanu, Dokskat, Kyango, Minaro*)	It is spoken in areas along the Indus river in Ladakh and Kargil districts in Northern Kashmir and in some areas such as Chulichan, Gurgurdo, Batalik and Dah.
74.	Bunan (*Ghara, Gahr*)	Spoken in some parts of Himachal Pradesh and Western Tibet
75.	Bundeli (*Bundelkhandi*)	It is spoken in Uttar Pradesh (Jalaun, Jhansi, Hamirpur districts), Madhya Pradesh (Balaghat, Chhindwara, Hoshangabad, Sagar, Sehore, Panna, Satna, Chhatarpur, Tikamgarh, Shivpuri, Guna, Bhind, Morena, Gwalior, Narsinghpur, Seoni, Datia districts), Maharashtra (Bhandara, Nagpur districts) and in some parts of Rajasthan, Gujarat and Andhra Pradesh. It has various dialects such as Bundeli, Pawari (Powari), Lodhanti (Rathora), Khatola, Banaphari, Kundri, Nibhatta, Bhadauri (Towargarhi), Lodhi, Koshti, Kumbhari, Nagpuri *Hindi* and Chhindwara Bundeli.
76.	Car Nicobarese	It is spoken in North Nicobar Islands.

S.N.	Languages/Dialect	Tribes and Distribution
77.	Chakma (*Takam, Chakama)*	It is spoken in some parts of Assam, Tripura, West Bengal, Mizoram and Manipur.
78.	Chamari (*Chamar, Chambhari*)	It is spoken in some parts of Madhya Pradesh, Uttar Pradesh and Maharashtra by the Chamar tribes.
79.	Chambeali (*Chamaya*)	It is spoken in some parts of Himachal Pradesh such as the Chamba district, and in parts of Jammu and Kashmir. *Bansbali*, *Bansyari* and *Gadi Chameali* are its dialects.
80.	Chaura (*Chowra, Tutet*)	It is spoken in Nicobar Islands and the Chaura Island.
81.	Chenchu (*Chenchucoolam, Chenchwar, Chenswar, Choncharu*)	It is a Dravidian language spoken in some parts of Andhra Pradesh, Tamil Nadu, Karnataka and Odisha.
82.	Chero	It is spoken in some parts of Bihar, Odisha, Uttar Pradesh and West Bengal.
83.	Chhattisgarhi (*Laria, Khaltahi*)	It is spoken in some parts of Madhya Pradesh, Bihar and Odisha. It has several dialects such as *Surgujia*, *Sadri Korwa*, *Baigani*, *Binjhwari*, *Kalanga*, *Bhulia*, *Chattisgarhi Proper*, *Kavardi* and *Khairagarhi*.
84.	Chik–Barik (*Chik Baraik, Panchpargania, Bedia, Pan, Pan Sawasi, Tanti*)	It is spoken in some parts of Bihar and West Bengal.
85.	Chin, Falam (*Halam Chin, Fallam, Tipura*)	It is spoken in some parts of Assam, Tripura, Mizoram and West Bengal, Myanmar and Bangladesh. Its dialects are *Chorei*, *Chari Chong*, *Halam*, *Kaipang*, *Kalai*, *Khelma*, *Mursum*, *Rupini* and *Shekasip*.
86.	Chin, Thado (*Thadou, Thado–Ubiphei, Thado–Pao, Kuki, Kuki–Thado*)	It is spoken in some parts Assam, Manipur, Nagaland and Mizoram. It has several dialects such as *Baite*, *Changsen*, *Jangshen*, *Kaokeep*, *Khongzai*, *Kipgen*, *Langiung*, *Sairang*, *Thangngen*, *Hawkip*, *Shithlou* and *Shingsol*.
87.	Chiru (*Chhori*)	It is spoken in some parts of Assam, Manipur and Nagaland.
88.	Chodri (*Chaudri*)	It is spoken mainly in Bharuch and Dang districts of Gujarat and in some parts of Maharashtra, Karnataka and Rajasthan.
89.	Chulikata (*Idu, Midhi, Midu, Ida*)	It is spoken in some parts of Assam, Arunachal Pradesh and West Bengal.
90.	Churahi (*Churahi Pahari*)	This is an Indo–Aryan language, which is spoken in parts of the Chamba district of Himachal Pradesh.
91.	Dal	It is spoken in some parts of Odisha.
92.	Darlong	It is spoken in some parts of Tripura and Bangladesh.
93.	Degaru	It is spoken in some parts of Bihar and West Bengal.

S.N.	Languages/Dialect	Tribes and Distribution
94.	Deori (*Chutiya, Drori*)	It is spoken in some parts of Assam, Arunachal Pradesh and Nagaland.
95.	Dhanki (*Dhanka, Dangri, Dangs Bhil, Tadavi, Tadvi Bhil, Kakachhu–Ki Boli*)	This is an Indo–Aryan language that is spoken in Dang district of Gujarat and in some parts of Maharashtra, Karnataka and Rajasthan.
96.	Dhanwar	It is spoken in some parts of Madhya Pradesh and Maharashtra.
97.	Dhodia (*Dhori, Dhowari, Doria*)	It is spoken in some parts of Gujarat, Daman and Diu, Dadra and Nagar Haveli, Madhya Pradesh, Maharashtra, Karnataka and Rajasthan.
98.	Digaro (*Digaru, Taraon, Taying, Mishmi*)	It is spoken in some parts of Assam and Arunachal Pradesh.
99.	Dimasa	It is spoken in North Cachar Hills and Cachar districts of Assam and parts of Nagaland. *Dimasa* and *Hariamba* are its dialects.
100.	Dogri–Kangri (*Dogri, Dhogaryali, Dogri Jammu, Dogri Pahari, Dongari, Tokkaru*)	It is an Indo–Aryan language, which is a mixture of Sanskrit, Punjabi and Persian. It is spoken by the Dogri tribes who are located in some parts of Jammu and Kashmir, West Bengal, Chandigarh and Himachal Pradesh (Kangra and Hamirpur districts). Its principal dialects are *Bhatbali*, *East Dogri*, *Kandiali*, *Kangri* (*Kangra*) and *North Dogri*.
101.	Dorli	This language is spoken by the *Dorli* tribes of the Bastar district of Madhya Pradesh.
102.	Dubla (*Dubli, Rathod, Talavia*)	This Indo–Aryan language is spoken in some parts of Gujarat, Maharashtra, Karnataka and Rajasthan.
103.	Duruwa (*Dhurwa, Parji, Thakara, Tagara, Tugara*)	It is a Dravidian language spoken mainly in South-east Jagdalpur, Madhya Pradesh (Bastar) and Odisha (Koraput). *Tiriya*, *Nethanar*, *Dharba* and *Kukanar* are its dialects.
104.	Gadaba (*Gadba, Gutob, Gudwa, Godwa, Bodo Gadaba, Boi Gadaba*)	It is spoken mainly in Andhra Pradesh (Visakhapatnam district) and Odisha (Koraput district). Dialects: *Munda Odisha Gadaba* and *Munda Andhra Pradesh Gadab*.
105.	Gadaba (*Ollar Gadaba, Ollari, San Gadaba, Gadba*)	It is spoken in some parts of Koraput district in Odisha.
106.	Gaddi (*Bharmauri Bhadi, Pahari Bharmauri, Gaddyaligadi*)	This Indo–Aryan language is spoken in some parts of Chamba district in Himachal Pradesh and parts of Uttar Pradesh and Jammu and Kashmir.
107.	Gade Lohar (*Gaduliya Lohar, Bagri Lohar, Lohari*)	This Indo–Aryan language is spoken in some parts of Rajasthan, Gujarat, Madhya Pradesh, Maharashtra, Uttar Pradesh, Delhi, Haryana and Punjab.

S.N.	Languages/Dialect	Tribes and Distribution
108.	Galong (*Galo, Adi–Gallong*)	It is spoken in parts of Assam and Arunachal Pradesh, in areas bordering Tibet.
109.	Gamti (*Gamati, Gamta*)	This Indo–Aryan language belonging to the *Bhil* sub-group is spoken by the *Gamti* tribes in parts of Bharuch and Surat in Gujarat.
110.	Gangte	It is spoken in some parts of Assam, Central Manipur, Meghalaya and Myanmar.
111.	Garhwali (*Gadhavali, Gadhawala, Gadwahi, Gorwali*)	This Indo–Aryan languge is spoken in some parts of Kashmir and Uttar Pradesh. Its various dialects are *Srinagaria*, *Tehri*, *Badhani*, *Dessaulya*, *Lohbya*, *Majh–Kumaiya*, *Bhattiani*, *Nagpuriya*, *Rathi* and *Salani*.
112.	Garo (Mande)	It is spoken in Western Assam, Garo Hills; Meghalaya, Nagaland, Tripura and West Bengal. It has various dialects like *Abengya*, *Achick*, *Ambeng*, *Atong*, *Awe*, *Chisak*, *Dacca*, *Ganching*, *Kamrup*, *Matchi*, *Megam* and *Ruga*.
113.	Gata (*Gataq, Geta, Didayi*)	Spoken in Koraput in Odisha
114.	Gawari (*Gamari*)	It is spoken in some parts of Andhra Pradesh, Madhya Pradesh, Bihar and Maharashtra.
115.	Girasia, Adivasi (*Adivasi Girasia, Garasia*)	This Indo–Aryan language belonging to the *Bhil* sub-group is spoken by the *Adivasi Girasia* tribes found in the Banaskantha and Sabarkantha districts of Northern Gujarat.
116.	Girasia, Rajput (*Rajput Garasia, Dungri Grasia, Dhungri Bhili*)	This Indo–Aryan language is spoken in Gujarat and Rajasthan.
117.	Gondi, Northern (*Gondi, Gondiva*)	This is a Dravidian language spoken mainly in Betul, Chindwara, Seoni, Mandla, and Balaghat districts of Madhya Pradesh and Amravati, Wardha, Nagpur, Bhandara, and Yavatmal districts of Maharashtra.
118.	Gondi, Southern (*Telugu Gondi*)	This is a Dravidian language spoken mainly in Adilabad district of Andhra Pradesh and Southern Yavatmal, Southern Chandrapur and Southeastern Gadchiroli districts of Maharashtra. Dialects: *Sironcha*, *Nirmal* (Adilabad), *Bhamragarh*, *Utnoor*, *Aheri*, *Rajura* and *Etapally Gondi*.
119.	Gowlan	This is an Indo–Aryan language spoken by *Korku* tribal people of Maharashtra.
120.	Gowli (*Nand*)	Indo–Aryan language spoken in Amravati district of Maharashtra and some parts of Madhya Pradesh. Dialects: *Nand*, *Ranya*, *Lingaayat* and *Khamla*.
121.	Groma (*Tromowa*)	Sino–Tibetan language spoken mainly in Sikkim

S.N.	Languages/Dialect	Tribes and Distribution
122.	Gujari (*Gujuri, Gojri, Gogri, Kashmir Gujuri*)	This Indo–Aryan language is spoken in some parts of Himachal Pradesh, Madhya Pradesh, Uttar Pradesh, Jammu and Kashmir and Rajasthan.
123.	Gurung (*Gurung Kura*)	This Sino–Tibetan language is spoken in parts of the Darjeeling district of West Bengal and in parts of Nepal.
124.	Hajong	This Indo–Aryan language is spoken in parts of Meghalaya, Assam, West Bengal and Bangladesh.
125.	Halbi (*Bastari, Halba, Halabi, Halvi, Mahari*)	This Indo–Aryan language is spoken in some parts of Madhya Pradesh (Bastar district), Maharashtra, Odisha and Andhra Pradesh. Dialects: *Adkuri, Bastari, Chandari, Gachikolo, Mehari, Muri* and *Sundi*.
126.	Harauti (*Hadauti, Piploda*)	Indo–Aryan language spoken in the Kota area of Rajasthan and some parts of Madhya Pradesh. *Sipari* and *Harauti* are its dialects.
127.	Haryanvi (*Bangaru, Bangri, Desari, Chamarwa*)	It is an Indo–Aryan language spoken by 13 million people or 85% of the population of Haryana. It is also spoken in some parts of Punjab, Delhi, Himachal Pradesh and Uttar Pradesh. Dialects: *Bangaru Proper, Deswali, Bagdi* and *Mewati*.
128.	Hmar (*Hamar, Mhar, Hmar)i*	It is spoken in some parts of Assam, Manipur and Mizoram.
129.	Ho (*Lanka Kol, Bihar Ho*)	This Austro–Asiatic language is spoken by over a million people belonging to the *Ho* tribes mainly in the Singhbhum district of Bihar and Mayurbhanj and Koenjhar districts of Odisha and parts of West Bengal and Bangladesh. Its important dialects are *Lohara* and *Chaibasa–Thakurmunda*. It is written in the Devanagari script in Bihar and the Oriya script in Odisha.
130.	Holiya (*Holar, Hole, Holu, Golari–Kannada*)	This is a Dravidian language spoken in some parts of Madhya Pradesh and Maharashtra.
131.	Hrangkhol (*Rangkhol*)	This is a Tibeto–Burman language spoken in parts of Manipur, Assam, Tripura and also in Myanmar. *Hadem* is its dialect.
132.	Ikrani	It is non-standard variety of Marathi typical of Dhule district in Maharashtra.
133.	Irula (*Eravallan, Erukala, Yerukala, Irava, Irular, Iruliga, Korava, Kad Chensu*)	This is a Dravidian language spoken mainly in Tamil Nadu (Nilgiri, Coimbatore, Periyar, Salem, and Chengai Anna), Karnataka, Kerala, Andhra Pradesh and Maharshtra. Dialects: *Mele Nadu, Vette Kada, Northern Irula* (*'Kasaba'*), *Bookapuram, Iruliga, Parikala* and *Sankara–Yerukala*.

S.N.	Languages/Dialect	Tribes and Distribution
134.	Jagannathi (*Jaga Aad, Surya Jagannathi*)	It is an Indo–Aryan language spoken in limited parts of Andhra Pradesh, Maharashtra and Karnataka. It is very similar to Oriya.
135.	Janggali (*Jhangal, Jhangar, Rawat*)	It is a Tibeto–Burman language with very limited speakers in the Pithoragarh district of Uttar Pradesh.
136.	Jangshung (*Jangrami, Zangram, Zhang–Zhung, Thebor, Thebör Skadd, Central Kinnauri*)	It is the Tibeto–Burman language spoken by about 4% of the population of Kinnaur district of Himachal Pradesh (1981 Census).
137.	Jarawa	This language is spoken in the Andaman Islands, interior and South-central Rutland Island, central interior and south interior South Andaman Island.
138.	Jatapu	This belongs to the Kui–Kuvi branch of Central Dravidian languages and is spoken mainly by some STs in the Srikakulam and Visakhapatnam districts of Andhra Pradesh, Ganjam and Koraput districts of Odisha and in some parts of Tamil Nadu.
139.	Jaunsari	This Indo–Aryan language is spoken in Jaunsar in Uttar Pradesh and parts of Dehradun in Himachal Pradesh.
140.	Jharia	This is an Indo–Aryan language spoken by a very limited number of people in the Koraput district of Odisha.
141.	Jingpho (*Chingpo, Kachin, Singpho, Chingpaw, Jinghpaw, Marip*)	Tibeto–Burman language, mainly spoken in Myanmar and China, but also spoken in some parts of Assam and Arunachal Pradesh.
142.	Juang (*Puttooas, Patua, Patra–Saara*)	This is an Austro–Asian language spoken in some parts of the Keonjhar district in Odisha.
143.	Juray	This is an Austro–Asian language spoken in some parts of Odisha.
144.	Kachari	This Tibeto–Burman language is spoken in the North Cachar Hills and Cachar districts of Assam.
145.	Kachchi (*Kutchchi, Katch*)	This Indo–Aryan language is spoken by nearly eight lakh people in India in parts of Andhra Pradesh, Madhya Pradesh, Uttar Pradesh, Assam, Gujarat (Kutch area), Kerala, Tamil Nadu, Maharashtra, Karnataka and Odisha. It is also spoken in Kenya, Tanzania and Malawi. *Jadeji* is the dialect of Kachchi.
146.	Kadar	This is a Dravidian language similar to Malayalam and is spoken by some STs in the Ernakulam, Palghat, and Trichur areas of Kerala; parts of Andhra Pradesh and Coimbatore district of Tamil Nadu.

S.N.	Languages/Dialect	Tribes and Distribution
147.	Kaikadi	This is a Dravidian language spoken by the *Kaikadi* tribes in some parts of Maharashtra and Karnataka.
148.	Kakborok (*Tripuri, Tripura, Kokborok*)	Kakborak is a Baric language belonging to the Tibeto–Burman language family. The phrase 'kok borok' literally means 'language people'. Kok borok is the mother tongue of a majority of the tribes in Tripura, and it is the language used in primary schools. It is spoken by over seven lakh people in India, mainly in Assam and eastern Tripura. It is also spoken in Bangladesh. Dialects: *Jamatia*, *Noatia*, *Riang* (*Tipra*), *Halam* and *Debbarma*. This language is also known as Usipi.
149.	Kamar	This is a Dravidian language spoken in some parts of Raipur and Rewa districts of Madhya Pradesh.
150.	Kanashi (*Kanasi*)	This language is spoken in parts of the Kulu valley in Himachal Pradesh.
151.	Kanauji (*Bhakha, Braj Kanauji*)	This is an Indo–Aryan language spoken in some parts of Uttar Pradesh. *Tirhari* is a dialect of Kanauji.
152.	Kanikkaran	This is a Dravidian language spoken in Kozikhode, Ernakulam and Thiruvanthapuram districts of Kerala and the Tirunelveli district of Tamil Nadu.
153.	Kanjari (*Kangar Bhat, Kangri*)	This Indo–Aryan language, sometimes called the 'Gypsy' language, is spoken by over 60,000 people in parts of Andhra Pradesh, Madhya Pradesh, Uttar Pradesh and Rajasthan. Dialect: *Kuchbandhi*.
154.	*Kannada*, Southern	It is a non-standard variety of *Kannada*, spoken by *Kurumba* people in Coimbatore and Dharmapuri districts, in pockets in Salem and North Arcot districts of Tamil Nadu and adjacent areas of Karnataka and Andhra Pradesh.
155.	Karmali (*Kohle*)	This Austro–Asiatic language is spoken in some parts of Bihar, Odisha and West Bengal.
156.	Katkari (*Katari, Kathodi, Katvadi*)	It belongs to the Indo–Aryan language family and is spoken by the Katkari tribes found in some parts of Gujarat and Maharashtra.
157.	Kawari	An Indo–Aryan language spoken by about 33,000 people in Raipur and surrounding districts of Madhya Pradesh and some parts of Maharashtra and Odisha.
158.	Keer (*Kir*)	It is in very limited use in some parts of Madhya Pradesh.
159.	Khamti (*Kham–Tai, Khantis, Tai Kam Ti*)	It is spoken by the *Khamt*i tribes found in some parts of Lakhimpur district of Assam and parts of Arunachal Pradesh.

S.N.	Languages/Dialect	Tribes and Distribution
160.	Khamyang	It is spoken in some parts of Assam.
161.	Khandesi (*Dhed Gujari*)	It is an Indo–Aryan language, which is spoken by the *Khandesi* tribes found in some parts of Maharashtra and Gujarat. Dialects: *Ahirani* (*Ahiri*), *Kunbi* (*Kunbau*), *Rangari*, *Khandesi* and *Kotali Bhil*.
162.	Kharia (*Haria, Kharvi, Khatria, Kheria, Khadia, Khariya*)	It is spoken primarily in the Ranchi district of Bihar and also in Raigarh district of Madhya Pradesh, Sundargarh district of Odisha and parts of Assam, Tripura, West Bengal and Andaman and Nicobar Islands. Dialects: *Dhelki Kharia*, *Dudh Kharia* and *Mirdha–Kharia*.
163.	Kharia Thar	It is spoken in some parts of Bihar.
164.	Khasi	It is an Austro–Asiatic language spoken in Khasi–Jaintia hills, Assam, Meghalaya, Jammu and Kashmir, Manipur, Punjab, Uttar Pradesh, and West Bengal. Dialects: *Bhoi–Khasi*, *Jaintia*, *Lyng–gam*, *Nongtung*, *Khasi* and *War*.
165.	Khirwar (*Khirwara*)	It is a Dravidian language spoken by the *Khirwar* tribes of the Sarguja district of Madhya Pradesh.
166.	Kinnauri, Chitkuli (*Chitkuli, Tsitkhuli, Kinnauri, Thebarskad*)	It is spoken in some parts of the Kinnaur district of Himachal Pradesh.
167.	Kinnauri, Lower (*Kinnaurayanuskad, Kanoreunu Skadd, Kanorug Skadd, Kunawur, Tibas Skad*)	It is spoken mainly in the Kinnaur district of Himachal Pradesh. It is also spoken in the Tibet border areas and in parts of Uttar Pradesh, Punjab, Kashmir and Himachal Pradesh. Dialects: *Harijan Kinnauri* and *Malhesti* (*Milchang*).
168.	Kinnauri, Upper (*Nyamskad, Mnyamskad, Myamkat, Nyamkat, Bud–Kat, Sangyas*)	It is spoken in some parts of the Kinnaur district of Himachal Pradesh. Dialects: *Nesang* and *Puh*.
169.	Kishanganjia (*Shreepuri, Siripuria*)	It is spoken in some parts of Bihar.
170.	Kishtwari	This is an Indo–Aryan language spoken by the Kashtwari tribes of Jammu and Kashmir.
171.	Koch (*Koc, Kocch, Koce*)	This is a Tibeto–Burman language spoken in some parts of Assam, Tripura, and Meghalaya, and also in Bangladesh. Dialects: *Banai*, *Harigaya*, *Satpariya*, *Tintekiya* and *Wanang*.
172.	Koch	This is an Indo–Aryan language spoken in the Cooch Behar district of West Bengal. It is distinct from Tibeto–Burman Koch spoken in Assam, Tripura and Bangladesh.

S.N.	Languages/Dialect	Tribes and Distribution
173.	Koda (*Kaora, Korali, Korati*)	It is spoken in some parts of Bihar, East Madhya Pradesh, Sambalpur district of Odisha and West Bengal. Dialects: *Khaira, Mirdha–Kora, Bankara, Birbhum* and *Dhangon*.
174.	Kodagu (*Kadagi, Khurgi, Kotagu*)	It is spoken mainly in the Kodagu (Coorg) district of Karnataka by the *Kodagu* tribes.
175.	Koireng	It is spoken in some parts of Manipur and Nagaland.
176.	Kol (*Dahit, Kolaboli, Kolar, Kolian, Kuli*)	It is spoken in some parts of Assam, Bihar, Madhya Pradesh, Maharashtra, Odisha and West Bengal.
177.	Kolai	It has very limited speakers in the border areas between Andhra Pradesh and Odisha.
178.	Kolami, North-western (*Kulme, Kolam, Kolmi*)	It is a Dravidian language that is spoken in some parts of Andhra Pradesh, Madhya Pradesh and Maharashtra (Yavatmal, Wardha, and Nanded districts). Dialects: *Madka–Kinwat, Pulgaon, Wani* and *Maregaon*.
179.	Kolami, Southeastern	It is a Dravidian language that is spoken in some parts of Adilabad district of Andhra Pradesh and Chandrapur and Nanded districts of Maharashtra. Dialects: *Metla–Kinwat, Utnur, Asifabad* and *Naiki*.
180.	Kom	It is a Tibeto–Burman language spoken in the east and central Manipur. *Kolhreng* is its dialect.
181.	Konda–Dora (*Porja*)	It is a Dravidian language that is spoken in some parts of Andhra Pradesh (Konda–Dora), Odisha (Kubi) and Assam. Dialects: *Konda* and *Kubi*.
182.	Konkani, Goanese (*Gomataki, Goan*)	It is an Indo–Aryan language that is spoken in the Southern coastal strip of Maharashtra, primarily in the districts of Ratnagari, and in Goa and in parts of Karnataka and Kerala. Dialects: Standard Konkani (*Goanese*), Bardeskari (*Gomantaki*), Sarasvat Brahmin, Kudali (*Malvani*), Daldi (*Nawaits*), Chitapavani (*Konkanasths*) and Mangalore.
183.	Koraga, Korra (*Koragar, Korangi*)	It is a Dravidian language that is spoken in some parts of Kerala.
184.	Koraga, Mudu	It is also a Dravidian language that is spoken in some parts of Kerala.
185.	Koraku	It is an Austro–Asiatic language spoken in parts of Surguja district of Madhya Pradesh.
186.	Korku (*Bondeya, Bopchi, Ramekhera, Kurku–Ruma*)	This language is spoken in parts of Southern Madhya Pradesh and Northern Maharashtra by the *Korku* or *Bapchi* tribes. Dialects: *Bouriya, Bondoy, Ruma* and *Mawasi*.

S.N.	Languages/Dialect	Tribes and Distribution
187.	Korwa	It is spoken in some parts of Bihar, Madhya Pradesh, Maharashtra, Uttar Pradesh and Andhra Pradesh. Dialect: *Majhi–Korwa*.
188.	Kota	It is a Dravidian language that is spoken in some parts of Chennai and Nilgiri Hills in Tamil Nadu. Dialect: *Ko Bashai*.
189.	Koya	It is a Dravidian language that is spoken in some parts of Andhra Pradesh, Maharashtra, Madhya Pradesh and Odisha. Dialects: *Malakanagiri Koya*, *Podia Koya* (*Gotte Koya*), *Chintoor Koya* (*Dorla Koitur*), *Jaganathapuram Koya* (*Gommu Koya*, *Godavari Koya*) and Dorli (*Korla*, *Dora*, *Dor Koi*, *Dora Koi*, *Dorla Koitur*, *Dorla Koya*).
190.	Kudiya	It is a Dravidian language that is spoken in some parts of Kannur district of Kerala and Kodagu (Coorg) and South Kannara districts of Karnataka and parts of Tamil Nadu.
191.	Kui	It is a Dravidian language that is spoken in some parts of Odisha, Andhra Pradesh, Madhya Pradesh and Tamil Nadu. Dialects: *Khondiand Gumsai*.
192.	Kukna (*Kanara, Kokna, Kokni*)	It is spoken in parts of Karnataka, Gujarat, Maharashtra and Dadra and Nagar Haveli.
193.	Kumarbhag Paharia (*Malto, Malti, Maltu*)	It is reported from the Santhal Pargana district, central eastern Bihar and Bankura, Bardhaman and Murshidabad districts of West Bengal.
194.	Kumauni (*Kumau, Kumman*)	It is spoken in some parts of Assam, Bihar, Delhi, Madhya Pradesh, Kumaon region of Uttar Pradesh, Maharashtra, Nagaland and Nepal. Dialects: *Askoti*, *Chaugarkhiya*, *Danpuriya*, *Gangola*, *Johari*, *Khasparjiya*, *Kumaiya* Pachhai, *Pashchimi*, *Phaldakotiya*, *Rau–Chaubhaisi* and *Soriyali*.
195.	Kupia	It is an Indo–Aryan language spoken by people called *Valmikis* in Andhra Pradesh.
196.	Kurichiya	It is spoken in parts of Kannur district of Kerala and parts of Tamil Nadu by the *Kurichiya* tribes.
197.	Kurumba	It is spoken in parts of Andhra Pradesh, Kerala, Tamil Nadu and Karnataka. Dialects: *Naik–Kuruba*, *Pal Kurumba*.
198.	Kurumba, Alu	It is a Dravidian language spoken in some parts of Tamil Nadu, especially along the eastern side of Nilgiri Hills.
199.	Kurumba, Betta	It is a Dravidian language spoken in some parts of Nilgiris district of Tamil Nadu; Mysore district of Karnataka and Waynad district of Kerala.

S.N.	Languages/Dialect	Tribes and Distribution
200.	Kurumba, Jennu (*Ten Kurumba*)	It is spoken on the northern side of Nilgiri Hills on the border between Tamil Nadu and Karnataka; Mysore and Kodagu districts of Karnataka and Waynad district of Kerala.
201.	Kurux (*Uraon, Kurukh, Oraon, Urang*)	It is a Dravidian language spoken in some parts of Bihar, Madhya Pradesh, Assam, Tripura, West Bengal and Odisha.
202.	Kuvi (*Jatapu*)	It is a Dravidian language spoken in some parts of Andhra Pradesh.
203.	Ladakhi	It is a Tibeto–Burman language spoken in the Leh and Ladakh areas of Jammu and Kashmir. Dialects: Leh (*Central Ladakhi*), Zangskari Ladakhi (*Zaskari*), Changtang Ladakhi (*Byangskat, Rong, Rupshu, Stotpa, Upper Ladakhi*), Shamma (*Shamskat, Lower Ladakhi*) and Nubra Ladakhi.
204.	Lahuli, Chamba (*Manchati, Manchad, Pattani, Chamba*)	It is a Tibeto–Burman language spoken in some parts of the Chamba district of Himachal Pradesh.
205.	Lahuli, Tinan (*Rangloi, Gondla, Tinan*)	It is spoken in the Lahaul and Spiti and Chanda Valleys of Himachal Pradesh.
206.	Lalung	It is a Sino–Tibetan language spoken in some parts of Assam and Meghalaya.
207.	Lamani (*Lambadi, Banjara, Gohar–Herkeri, Goola, Gurmarti, Kora, Labhani Lavani, Vanjari, Wanji, Singali*)	It is spoken by the *Banjaras* in some parts of Andhra Pradesh, Madhya Pradesh, Himachal Pradesh, Gujarat, Tamil Nadu, Maharashtra, Karnataka, Odisha and West Bengal.
208.	Lamkang	It is spoken in some parts of South-east Manipur and Nagaland.
209.	Lepcha (*Rongke, Rongpa, Nunpa*)	It is spoken in Nepal; Bhutan; Sikkim and West Bengal. Dialects: *Ilammu*, *Tamsangmu* and *Rengjongmu*.
210.	Lhoba, Yidu (*Luoba, Lho–Pa*)	It is mainly spoken in the East and West Kamang districts of Arunachal Pradesh and in parts of China.
211.	Lhomi (*Lhoket, Shing Saapa*)	It is spoken in the Darjeeling district of West Bengal and also in Nepal and China.
212.	Limbu	It is spoken in the Darjeeling district of West Bengal and in parts of Sikkim.
213.	Lodhi (*Lodha*)	It is spoken in part of Bihar, Madhya Pradesh, Maharashtra, Odisha and West Bengal.
214.	Lushai (*Dulien, Lukhai, Lusago, Sailau, Whelngo*)	It is spoken in some parts of Mizoram, Assam, Manipur, Nagaland, and Tripura and in Bangladesh and Myanmar. Dialects: *Fannai*, *Mizo*, *Ngente* and *Pang* (*Paang*).

S.N.	Languages/Dialect	Tribes and Distribution
215.	Magahi (*Magadhi, Magaya, Magi*)	It is spoken in the Southern districts of Bihar and Malda district of West Bengal.
216.	Magar, Eastern (*Magari*)	It is spoken in Nepal, Sikkim and Bhutan.
217.	Mahl	It is spoken in Maldives and in the Minicoy islands of Laskshadweep.
218.	Mahali	It is spoken in some parts of Assam, Bihar, Odisha and West Bengal.
219.	Maithili	It is spoken by over 22 million people in India spread over Bihar, Delhi, Odisha, Madhya Pradesh, Maharashtra and West Bengal. Dialects: *Standard Maithili, Southern Standard Maithili, Eastern Maithili* (*Khotta, Kortha*), *Chikachiki, Western Maithili, Jolaha, Central Colloquial Maithili* (*Sotipura*), *Kisan* and *Angika*.
220.	Maithili, Dehati (*Dehati, Dahati, Deshiya, Deshia*)	This language is different from standard Maithili and is spoken in Bihar, Odisha and some districts of South Nepal bordering Bihar.
221.	Majhi	It is spoken in parts of Gurdaspur and Amritsar districts of Punjab and also in Lahore.
222.	Majhwar	It is spoken in some parts of Madhya Pradesh, Uttar Pradesh and Sikkim.
223.	Mal Paharia (*Malto, Malti, Maler, Malpaharia, Maw, Mawdo*)	It is spoken in some parts of Bihar and West Bengal.
224.	Malankuravan	It is a Dravidian language spoken in some parts of Kerala.
225.	Malapandaram	It is a Dravidian language spoken in some parts of Kottayam, Ernakulam and Quilon areas of Kerala.
226.	Malaryan (*Maleyarayan, Male Arayans, Arayans, Karingal, Vazhiyammar*)	It is a Dravidian language spoken in some parts of Ernakulam, Kottayam and Trissur districts of Kerala and parts of Tamil Nadu.
227.	Malavedan	It is a Dravidian language spoken in some parts of Ernakulam, Kottayam, Thiruvanthapuram districts of Kerala and Tirunelveli district of Tamil Nadu. Dialects: *Vetan* and *Vettuvan*.
228.	Maldivian (*Malikh, Mahl, Malki, Divehi, Divehli*)	It is the main language Maldives, but a few thousand speakers are found on the Minicoy Island in the Laccadive Islands in India.

S.N.	Languages/Dialect	Tribes and Distribution
229.	Mali	It is spoken in some parts of Andhra Pradesh and Maharashtra.
230.	Malto	This language is spoken by the Malto tribes found in the Rajmahal Hills of Bihar and West Bengal.
231.	Malvi (*Malwada, Mallow, Ujjaini, Malwi, Malavi*)	This Indo–Aryan language is spoken in some parts of North-west Madhya Pradesh, Maharashtra, Rajasthan and Gujarat. Dialects: *Bachadi*, *Bhoyari*, *Dholewari*, *Hoshangabad*, *Jamral*, *Katiyai*, *Malvi Proper*, *Patvi*, *Rangari*, *Rangri* and *Sondwari*.
232.	Manda	It is a Dravidian language spoken in some parts of the Kalahandi district of Odisha.
233.	Mandeali (*Mandi*)	It is spoken in the Mandi district of Himachal Pradesh.
234.	Mangelas	It is similar to *Konkani* and is spoken in some parts of Maharashtra.
235.	Manjhi	This language is spoken by the *Manjhi* tribes found in the Gumla district of Bihar.
236.	Manna–Dora	It is a Dravidian language spoken in some parts of East Godavari, Srikakulam and Visakhapatnam districts of Andhra Pradesh and parts of Tamil Nadu.
237.	Mannan (*Manne, Mannyod*)	It is a Dravidian language spoken in some parts of Kerala and Andhra Pradesh.
238.	Maria (*Madi, Madiya, Modi*)	It is a Dravidian language spoken in some parts of the Chanda district of Maharashtra and some parts of Madhya Pradesh. Dialects: *Bhamani Maria* (*Bhamani*), *Adewada* and *Etapally Maria*.
239.	Maria, Dandami (*Bison Horn Maria, Maria Gond, Madiya, Dhuru, Dandami Madiya*)	It is a Dravidian language spoken in central and Southern Bastar district of Madhya Pradesh. Dialects: *Geedam* and *Sukma* (*Suka*).
240.	Marwari (*Rajasthani, Mewari*)	It is spoken by over 12 million people in Gujarat, Rajasthan, Madhya Pradesh, Punjab, Delhi, Haryana and Uttar Pradesh.
241.	Matia	Spoken by *Adivasis* of Andhra Pradesh.
242.	Mawchi	The Mavchi or Mawchi language belongs to the *Bhil* branch of the Indo–Aryan language family. It is spoken in South-west Gujarat and Maharashtra by the *Mavchi* tribes.
243.	Miju (*Kaman, Mishmi, Miji*)	It is a Tibeto–Burman language spoken in some parts of Assam and Arunachal Pradesh.

S.N.	Languages/Dialect	Tribes and Distribution
244.	Mikir (*Manchati, Mikiri, Karbi*)	It is spoken in Mikir and Rengma Hills of Assam; Arunachal Pradesh, Meghalaya and Nagaland. Dialects: *Amri, Arleng, Bhui* and *Rhengkitang.*
245.	Mina	It is spoken in some parts of Madhya Pradesh and Rajasthan.
246.	Mirdha	It is an Austro–Asian language spoken in some parts of Odisha.
247.	Mirgan (*Panika, Panka*)	It is spoken mainly in the Bastar district of Madhya Pradesh and Koraput district of Odisha.
248.	Moinba (*Monba, Mompa*)	It is a Tibeto–Burman language spoken in Arunachal Pradesh, Bhutan and China.
249.	Mru (*Mro, Murung, Niopreng*)	It is spoken by the *Mru* tribes in some parts of West Bengal; Myanmar and Bangladesh.
250.	Mukha–Dora (*Reddi–Dora, Conta–Reddi, Reddi*)	It is spoken in some parts of Andhra Pradesh and Tamil Nadu.
251.	Mulia	It has limited number of speakers in Andhra Pradesh.
252.	Mundari (*Mandari, Munda, Horo*)	It has over a million speakers in Assam, Southern and Western parts of Ranchi district in Bihar and some areas in Himachal Pradesh, Madhya Pradesh, Odisha, Tripura, West Bengal and Andaman and Nicobar Islands.
253.	Muria, Eastern	It is a Dravidian language spoken in Northeastern Bastar district of Madhya Pradesh and Northwestern Koraput district of Odisha. Dialects: *Raigarh* and *Lanjoda.*
254.	Muria, Western (*Jhoria, Mudia, Muria Gondi*)	It is spoken mainly in the northern and western Bastar district of Madhya Pradesh. Dialects: *Sonapal*, *Banchapai* and *Dhanora.*
255.	Muthuvan	It is a Dravidian language spoken in Kozhikode, Kannur, Ernakulam, Kottayam and Trissur districts of Kerala and parts of Andhra Pradesh.
256.	Naga Pidgin (*Nagamese*)	It is spoken in Nagaland, especially in the Kohima district.
257.	Nagarchal (*Nagar, Nagarchi*)	It is a Dravidian language spoken in some parts of Madhya Pradesh, Maharashtra and Rajasthan.
258.	Naga, Angami (*Ngami, Angamis, Tsoghami, Monr, Tsanglo)*	It is spoken in parts of Western Nagaland, Manipur and Maharashtra. Dialects: *Kohima, Dzuna, Kehena, Khonoma, Chakroma* (*Western Angami*), *Mima, Nali, Mozome, Tengima* and *Tenyidie* (*Tenyidye*).
259.	Naga, Ao (*Aorr, Paimi, Cholimi, Nowgong, Hatigoria, Uri, Ao*)	It is spoken in Northeastern Nagaland and parts of Assam. Dialects: *Mongsen Khari, Changki, Chongli* (*Chungli*), *Dordar* (*Yacham*) and *Longla.*

S.N.	Languages/Dialect	Tribes and Distribution
260.	Naga, Chang (*Mojung, Machongrr, Mochumi, Changyanguh*)	It is spoken in some parts of Assam and in east central Nagaland.
261.	Naga, Chokri (*Eastern Angami, Chakrima Naga, Chakru, Chakhesang*)	It is spoken in some parts of Nagaland.
262.	Naga, Chothe	It is spoken in South-east Manipur and parts of Nagaland.
263.	Naga, Kabui (*Kapwi, Kubai*)	It is spoken in Manipur, Nagaland and Assam.
264.	Naga, Khezha (*Kezami, Khezhama, Khezha*)	It is spoken in parts of the Kohima district of Nagaland.
265.	Naga, Khiamngan (*Khiamngan, Makware, Nokaw, Para, Ponyo, Aoshedd, Welam)*	It is spoken in the east central part of Tuensang district in Nagaland.
266.	Naga, Khoibu Maring	It is spoken in some parts of Manipur.
267.	Naga, Khoirao (*Kolya, Mayangkhang, Thanggal, Tukaimi*)	It is spoken in parts of North Manipur.
268.	Naga, Konyak (*Kanyak, Konyak)*	It is spoken in some parts of Assam and the Mon District of Nagaland. Dialects: *Angphang*, *Hopao*, *Changnyu*, *Chen*, *Chingkao*, *Chinglang*, *Choha*, *Gelekidoria*, *Jakphang*, *Kongon*, *Longching*, *Longkhai*, *Longmein*, *Longwa*, *Mohung*, *Tableng*, *Mon*, *Mulung*, *Ngangching*, *Sang*, *Shanlang*, *Shunyuo*, *Shengha*, *Sima*, *Sowa*, *Shamnyuyanga*, *Tabu*, *Tamkhungnyuo*, *Tang*, *Tobunyuo*, *Tolamleinyua* and *Totok*.
269.	Naga, Liangmai (*Liyang, Lyangmay*)	It is spoken in some parts of Nagaland.
270.	Naga, Lotha (Chizima, Miklai, Tsontsii	It is spoken in some parts of Nagaland.
271.	Naga, Mao (*Mao, Spowama, Sopvoma, Sopfomo*)	It is spoken in some parts of North-west Manipur and Nagaland.
272.	Naga, Maram (*Maram*)	It is spoken in some parts of Assam and North Manipur
273.	Naga, Maring (*Maring*)	It is spoken in some parts of Manipur.
274.	Naga, Meluri (*Meluri, Meluory, Anyo*)	It is spoken in some parts of Nagaland.
275.	Naga, Monsang:	It is spoken in some parts of Manipur.

S.N.	Languages/Dialect	Tribes and Distribution
276.	Naga, Moyon (*Mayol*)	It is spoken in some parts of Nagaland and Manipur.
277.	Naga, Mzieme	It is spoken of Southwestern Nagaland.
278.	Naga, Nocte (*Borduria, Jaipuria, Mohongia, Namsangia*)	It is spoken in some parts of Nagaland, Arunachal Pradesh and Assam.
279.	Naga, Ntenyi	It is spoken in Westcentral Nagaland.
280.	Naga, Phom (*Tamlu Naga, Chingmengu*)	It is spoken in some parts of Nagaland.
281.	Naga, Pochuri	It is spoken in some parts of South-east Nagaland.
282.	Naga, Poumei	It is spoken in some parts of Manipur.
283.	Naga, Rengma (*Rengma, Mozhumi, Moiyui, Mon, Unza, Nzong, Nzonyu, Injang*)	It is spoken in West central Nagaland, Assam and Manipur. Dialects: *Keteneneyu* and *Azonyu*.
284.	Naga, Rongmei (*Maruongmai, Nruanghmei, Rongmei*)	It is spoken in North-west Manipur, Nagaland and Assam.
285.	Naga, Tarao	It is spoken in parts of South-east Manipur.
286.	Naga, Tase (*Tasey, Tangsa, Rangpan, Cham Chang*)	It is spoken in northern Nagaland and Myanmar. Dialects: *Longphi*, *Yogli*, *Have*, *Khemsing*, *Lungchang*, *Lungri*, *Moklum*, *Ponthai*, *Rongrang*, *Ronrang*, *Taipi*, *Tikhak*, *Sanke* (*Shangge*) and *Sangche*.
287.	Naga, Wancho (*Wancho, Banpara Naga, Joboka*)	It is spoken in some parts of Assam, Nagaland and Arunachal Pradesh. Dialects: *Changnoi*, *Bor Muthun*, *Horu Muthun* and *Kulung Muthun*.
288.	Naga, Yimchungru	It is spoken in some parts of Assam and Nagaland Dialects: *Tikhir*, *Wai*, *Chirr*, *Minir*, *Pherrongre* and *Yimchungru*.
289.	Naga, Zeme (*Kachcha, Mezama, Sangrima, Sengima, Arung, Empui, Jeme, Zemi*)	It is spoken in some parts of Manipur, Nagaland and Assam. Dialects: *Paren* and *Njauna*.
290.	Nahari	It is spoken in Raipur, Bilaspur districts of Madhya Pradesh and Sambalpur district of Odisha.
291.	Newari	It is spoken in some parts of Bihar.
292.	Nicobarese, Central (*Nicobar*)	It is spoken in Nicobar Islands, Katchall, Camorta, Nancowry and Trinkat islands. Dialects: *Camorta*, *Katchal*, *Nancowry* and *Trinkut*.

S.N.	Languages/Dialect	Tribes and Distribution
293.	Nicobarese, Southern	It is spoken in Little Nicobar and outer Great Nicobar islands. Dialects: *Condul*, *Great Nicobar*, *Little Nicobar* and *Milo*.
294.	Nihali	It is spoken in some parts of Madhya Pradesh and Maharashtra.
295.	Nimadi	It is spoken in some parts of Madhya Pradesh, Uttar Pradesh and Maharashtra. Dialect: *Bhuani*.
296.	Nisi (*Dafla, Nishi, Nishang, Bangni, Lel*)	This Tibeto–Burman language is spoken in some parts of Assam and Arunachal Pradesh. Dialect: *Aka Lel*.
297.	Nuka–Dora	It is Dravidian language spoken by *Adivasis* in Andhra Pradesh and Odisha.
298.	Ojhi (*Ojha*)	It is spoken in some parts of Madhya Pradesh and Maharashtra.
299.	Oko–Juwoi	It is spoken in parts of Andaman Islands.
300.	Ollari	It is a Dravidian language spoken mainly in Koraput in Odisha.
301.	Onge	It is spoken in Southern Andaman Islands.
302.	Oriya, Adivasi	It is spoken by *Adivasis* in Araku Valley in Andhra Pradesh and parts Odisha. Dialect: *Valmiki Adivasi Oriya*.
303.	Pahari, Kullu	It is spoken in parts of Himachal Pradesh and Punjab. Dialects: *Inner Siragi Kukui* and *Sainji*.
304.	Pahari, Mahasu	It is spoken in some parts of Shimla and Solan districts of Himachal Pradesh. Dialects: Lower Mahasu Pahari (*Kiunthali, Sirmauri, Baghati, Handuri, Baghliani*) and Upper Mahasu Pahari (*Shimla Siraji, Sodochi, Rampuri, Rohruri*).
305.	Paliyan (*Palaya, Seramar*)	It is a Dravidian language spoken in some parts of Kerala and Tamil Nadu.
306.	Pangi	It is spoken mainly in Himachal Pradesh in the Lahul–Spiti and Chamba districts.
307.	Panika	It is spoken in the Shahdol district of Madhya Pradesh.
308.	Paniya	It is a Dravidian language spoken in some parts of Kerala and Tamil Nadu.
309.	Pao (*Pabra*)	It is spoken in the Satna district of Madhya Pradesh.
310.	Pardhan	It is a Dravidian language spoken in some parts of Andhra Pradesh, Madhya Pradesh and Maharashtra.
311.	Pardhi	It is a Dravidian language spoken in some parts of Andhra Pradesh, Madhya Pradesh, Gujarat and Maharashtra. Dialects: *Neelishikari*, *Pittala Bhasha* and *Takari*.

S.N.	Languages/Dialect	Tribes and Distribution
312.	Parenga (*Parengi, Gorum*)	It is spoken in the Koraput district of Odisha and parts of Andhra Pradesh.
313.	Patelia	It is spoken in some parts of Gujarat and Maharashtra.
314.	Pauri	It is spoken in some parts of Madhya Pradesh and Maharashtra. *Rath* is its dialect.
315.	Pengo	It is a Dravidian language spoken in parts of Koraput district of Odisha.
316.	Phake	It is spoken in some parts of Assam.
317.	Phudagi	It is spoken in some parts of Maharashtra.
318.	Pnar (*Synteng*)	It is spoken in some parts of Khasi and Jaintia Hills of Meghalaya. Dialects: *Jaintia* and *Nongtung*.
319.	Puh	It is spoken in the Kinnaur district of Himachal Pradesh.
320.	Punjabi, Western (*Lahnda, Lahanda, Lahndi*)	It is spoken by over 40 million people in Pakistan and in parts of Jammu and Kashmir, Delhi and Haryana.
321.	Purik (*Burig, Burigskat*)	It is a Tibeto–Burman language, closely related to Balti, and dominant in the Kargil district in North Kashmir.
322.	Rabha (*Rava*)	It is Tibeto–Burman language spoken in West Assam, Garo Hills, Nagaland, West Bengal and Meghalaya. Dialects: *Maitaria, Rangdania*.
323.	Rajbangsi (*Tajpuri*)	It is an Indo–Aryan language spoken in parts of Jalpaiguri, Cooch behar, Darjeeling districts of West Bengal and also in Nepal and Bangladesh. Dialect: *Bahe*.
324.	Ralte	Spoken in Myanmar, but also in some parts of Assam
325.	Rawang	This language is part of the Sino–Tibetan language family and is spoken by the *Kiutze* tribes located mostly near the Myanmar–Tibet border.
326.	Reli	Indo–Aryan language spoken in some parts of Andhra Pradesh and Odisha.
327.	Riang (*Kau Bru*)	It is a Tibeto–Burman language spoken in some parts of Assam, Central Tripura, Mizoram and Bangladesh.
328.	Sadri (*Sadani, Santri, Siddri, Nagpuria, Chota Nagpuri, Dikku Kaji*)	It spoken in some parts of Assam, Bihar, Madhya Pradesh, West Bengal, Maharashtra, Odisha, Andaman Islands and Nagaland.
329.	Saharia (*Sor, Sosia*)	It is spoken in some parts of Madhya Pradesh.
330.	Samvedi	It is spoken in some parts of Maharashtra.
331.	Santali (*Hor, Har, Satar, Sandal, Sangtal,*	It is an important *adivasi* language belonging to the Austric family. It is spoken by over 5 million people spread over in Bihar, Odisha, West Bengal, Tripura and Assam. It is also

S.N.	Languages/Dialect	Tribes and Distribution
	Santhali, Sentali)	spoken in Bhutan and Bangladesh. Dialects: *Karmali* (*Khole*), *Kamari–Santali*, *Lohari–Santali*, *Mahali* (*Mahle*), *Manjhi* and *Paharia*. It does not have a script, but is written in *Bengali*, *Oriya*, *Nagari* or Roman scripts.
332.	Saraiki (*Multani*, Southern *Punjabi, Reasati*)	It is spoken by over 15 million people in Pakistan, but a few thousand speakers are found in India in some parts of Punjab, Maharashtra, Andhra Pradesh, Madhya Pradesh, Uttar Pradesh, Rajasthan, Delhi and Gujarat. Dialects: *Jafri*, *Siraiki Hindki*, *Thali* and *Jatki*.
333.	Saurashtri (*Patnuli*)	Indo–Aryan language spoken in some parts of Chennai, Madurai, Thanjavur, Tiruchchirappalli, Tirunelveli, North Arcot and Salem cities of Tamil Nadu and some parts of Karnataka and Andhra Pradesh.
334.	Sauria Paharia (*Malto, Malti*)	Dravidian language spoken in some parts of Santhal Pargana and Rajmahal hills of Bihar and Bankura, Bardhaman and Murshidabad districts of West Bengal. Dialects: *Sahibganj*, *Godda*, *Hiranpur* and *Litipara* (*Chatgam*).
335.	Savara	It is a Dravidian language spoken in some parts of Andhra Pradesh.
336.	Sentinel (*Sentinelese*)	It is spoken in Sentinel Island and other Southeastern Andaman Islands.
337.	Shendu (*Khyen, Khieng*)	It is a Tibeto–Burman language spoken in Lushai Hills of Assam and in Bangladesh.
338.	Sherdukpen	It is spoken in some parts of Assam and Arunachal Pradesh.
339.	Sherpa (*Sharpa Bhotia, Xiaerba, Serwa*)	Tibeto–Burman language spoken in the Darjeeling district of West Bengal, Sikkim and Arunachal Pradesh.
340.	Shina (*Shinaki*)	It is spoken in the Dras Valley and Gurais area in Kishenganga Valley in Northern Kashmir. Dialects: *Drasi* and *Gurezi*.
341.	Shobang	It is spoken in some parts of Andaman and Nicobar Islands.
342.	Sholaga	Dravidian language spoken in some parts of Tamil Nadu and Karnataka
343.	Shom Peng	It is an Austro–Asiatic language spoken in the Great Nicobar Island.
344.	Shrin	This language is spoken by the *Dog–Pa* or *Shrin* tribes of Jammu and Kashmir.
345.	Shumcho (*Thebor, Central Kinnauri*)	Tibeto–Burman language spoken in some parts of the Kinnaur district of Himachal Pradesh.
346.	Sikkimese (*Sikkim Bhotia, Danjongka*)	Tibeto–Burman language spoken in northern half of Sikkim.

S.N.	Languages/Dialect	Tribes and Distribution
347.	Simte	Tibeto–Burman language spoken in South-west Manipur.
348.	Sondwari	Indo–Aryan language spoken in some parts of Madhya Pradesh and Rajasthan.
349.	Sora (*Shabari, Savara, Sawaria*)	Austro–Asiatic language spoken mainly in the Ganjam, Koraput and Phulbani areas of Odisha and also in some parts of Andhra Pradesh, Madhya Pradesh, Bihar, Tamil Nadu, West Bengal and Assam.
350.	Sunam	It is spoken in some parts of the Kinnaur district of Himachal Pradesh.
351.	Tamang, Eastern	It is spoken in Darjeeling in West Bengal and in parts of Sikkim.
352.	Tamaria (*Temoral*)	It is spoken in Ranchi and Singhbhum districts of Bihar.
353.	Teressa (*Taih–Long*)	It is spoken in Nicobar Islands, Teressa and Bompoka islands. Dialect: *Bompoka.*
354.	Thakuri	Indo–Aryan language spoken in some parts of Maharashtra and Odisha.
355.	Tharu, Buksa (*Buksa*)	Indo–Aryan language spoken in some parts of Nainital, Bijnor and Garhwal districts of Uttar Pradesh. Dialects: *Madnapur* and *Thari.*
356.	Tibetan (*Lhasa, Bhotia,Dalai, Pohbetian*)	Spoken mainly in Tibet, Nepal, and Bhutan and in India in Himachal Pradesh, Arunachal Pradesh, Assam, Delhi and Sikkim. Dialects: *Kongbo, Spitiand Stod.*
357.	Toda (*Todi*)	Dravidian language spoken in parts of Odisha and Tamil Nadu.
358.	Tsangla	It is spoken mainly in Bhutan, but is also spoken in several villages in Arunachal Pradesh.
359.	Tulu	It is a Dravidian language spoken mainly in the Udipi and South Kannara districts and Managalore in Karnataka. It is also spoken by few people in Andhra Pradesh, Kerala, Tamil Nadu and Maharashtra. It does not have a separate script. A large number of Yakshaganas are composed in Tulu.
360.	Turi	It is spoken in some parts of Madhya Pradesh and Odisha.
361.	Ullatan (*Katan, Kattalan, Kochuvelan*)	Dravidian language spoken in some parts of Kerala
362.	Urali (*Oorazhi*)	It is spoken in some parts of Periyar and Nilgiri areas of Tamil Nadu.
363.	Vadval (*Phudagi*)	It is spoken in some parts of Maharashtra.
364.	Vaiphei	It is spoken in some parts of Assam, Manipur and Meghalaya.

S.N.	Languages/Dialect	Tribes and Distribution
365.	Varhadi–Nagpuri (*Berari, Dhanagari, Kumbhari)*	It is spoken in some parts of Maharashtra. Dialects: *Brahmani, Kunbi, Raipur, Jhadpi, Govari, Kosti* (*Rangari*), *Kunban* (*Kohli*) and *Mahari* (*Dhedi*).
366.	*Warli*	It is spoken in some parts Maharashtra and Gujarat.
367.	Vasavi	It is spoken in some parts Maharashtra and Gujarat.
368.	Vishavan (*Malarkuti*)	Dravidian language spoken in some parts of Ernakulam, Kottayam and Trichur districts of Kerala.
369.	Waddar	Dravidian language spoken in some parts of Andhra Pradesh, Karnataka and Maharashtra.
370.	Wagdi (*Wagri, Wagholi, Mina Bhil*)	It is spoken in some parts of Udaipur, Dungarpur and Banswara regions of Rajasthan and in Sabarkantha and Panch Mahals regions of Gujarat. Dialects: *Kherwara, Sagwara* and *Adivasi*.
371.	Yanadi (*Yadi, Yandis*)	Dravidian language spoken in some parts of Andhra Pradesh, Tamil Nadu and Odisha.
372.	Yerava	Dravidian language spoken mainly in the Kodagu district of Karnataka.
373.	Zanskari	It is a Tibeto–Burman language spoken along the Zaskar mountains and in the Kargil valley of Kashmir.
374.	Zome	It is spoken mainly in Myanmar, but is also spoken in some parts of Manipur and Assam.

Source: Culturopedia, 2014

BIBLIOGRAPHY

Adivasi Wikipedia: The Free Encyclopedia 2014 *http://en.wikipedia.org*. Retrived on December 08, 2014

Bahadur KP 1997 Caste Tribes and Culture of India: Bengal, Bihar and Orissa, New Delhi: ESS Publications

Bakshi SR and Kiran B 2000 *Social and Economic Development of Scheduled Tribes*, New Delhi: Deep and Deep Publication Pvt. Ltd.

Baradwaj AN 1985 Problems of Scheduled Caste and Scheduled Tribes in India, New Delhi: Light and Life Publishers

Bhatnagar SP, Banerjee NK, and Roy Burman BK 1969 Bibliography of Publications in Tribal Languages, Vol-45, No. 3, Linguistic Society of America

Census of India 2011 http://censusindia.gov.in/. Retrieved on September 11, 2014 Ministry of Home Affairs, Government of India, New Delhi-

Census of India 2011 Primary Census Abstract- Data Highlight, *http://www.censusindia.gov.in*. Retrieved on November 06, 2014Registrar General & Census Commissioner, Ministry of Home Affairs

Channa CM (ed) 2004 *Encyclopaedia of Indian Tribes and Castes*, New Delhi: Cosmo Publications

Chattopadhyaya K 1978 *Tribalism in India*, New Delhi: Vikas Publishing house Pvt. Ltd.

Culturopedia: 2014 Languages in India, Minor and tribal Languages, *http://www.culturopedia.com*. Retrieved on December 06, 2014

Culturopedia: 2014 Tribes of India, *http://www.culturopedia.net*. Retrieved on October 10, 2014

Das TC 1973 Social organisation, in The Tribal People of India, New Delhi: Publication Division, Ministry of Information and Broadcasting, Government of India

Devath S 2014 Tribal Development through Five Year Plans in India - An overview, *The Dawn Journal* 3 794–816

Elwin V 1944 *The Muria and Their Ghotul,* London: Oxford University Press

Ember CR and Ember M 1977 *Cultural Anthropology*, Englewood Cliffs: Prentice-Hall

Fernandez W and Menon G 1987 Tribal Women and Forest Economy, Afforestation, Exploitation and Status Change, New Delhi: Indian Social Institute

Fernandez W, Menon G, and Viegas P 1988 Forest, Environment and Tribal Economy, Deforestation, Improvement avid Marginalisation in Orissa, New Delhi: Indian Social Institute, Tribes of India Series

Forest Survey of India, Ministry of Environment and Forest, Government of India, (2013): India State Forest Report - 2013

Foster GM 1967 *Mexican Peasants in a Changing World*, Boston: Little, Brown

Gare GM and Aphale MB 1982 The Tribes of Maharashtra, Pune: Tribal Research and Training Institute

Harris M 1987 *Cultural Anthropology*, New York: Harper and Row Publishers

Haseena VA 2014 Land Alienation and Livelihood Problems of Scheduled Tribes in Kerala, Research on Humanities and Social Sciences 4 76–81

Iyer C and Sriya FM 2002 *Demography and Religion in India*, New Delhi: Oxford University Press

Jain AK and Sharma AN 2012 Impact of Socio-religious Believes and Taboo Practices on Health Status among the Baigas, *International Journal of Behavioral Social and Movement Sciences*, 1 1–8

Joshua Project 2015, Map Resources, http://joshuaproject.net/map_navigation, retrieved on May 06, 2015

Kamlesh M 1996 *Tribal Women: On the threshold of Twenty-First Century*, New Delhi: MD Publication Private Limited

Karmarkar KG 2002 The Silenced Drums – A Review of Tribal Economic Development, New Delhi: Northern Book Centre

Kroeber AL 1956 The Place of Boas in Anthropology, *American Anthropologist* 58 151–159

MacKenzie P 2009 Mother tongue first multilingual education among the tribal communities in India, *International Journal of Bilingual Education and Bilingualism* 12 369–385

Madan TN 2001 A Family and kinship: a study of the Pandits of Rural Kashmir, New Delhi: Oxford University Press

Mahalingam S 1990 Role of Co-operative Organisations in Developing Tribal Economy, New Delhi: Mittal Publications

Maharashtra Forest Department, Government of Maharashtra, (2013): A Statistical Outline, Current Salient Forest Statistics, 2013

Maheswari JK 1990 Interaction of Tribal with the Forests, Paper presented in tropical forest ecosystem conservation and development in south and South-East Asis: Proceeeding MAB Regional Training Workshop, KFRI, Peechi, Kerala Page 115-126

Majumdar DN 1961 *Races and Cultures of India*, Bombay: Asia Publishing House

Majumdar DN and Madan A 1989 *An Introductuon to Social Anthropology*, New Delhi: National Publishing House

Mamoria CB 1958 Tribal Demography in India, Allahabad: Kitab Mahal

Maps of India 2015, Indian languages, http://www.mapsofindia.com/culture/indian-languages.html, retrieved on April 12, 2015

Ministry of Tribal Affairs, Government of India, (2012): Demographic Status of Scheduled Tribe Population of India

Ministry of Tribal Affairs, Government of India, (2012-13) and (2013-14): Annual Report

Ministry of Tribal Affairs, Government of India, (2014): Tribal Profile at a Glance (May-2014)

Ministry of Tribal Affairs, New Delhi: 2004 *The National Tribal Policy (draft),* http:litribal.nic.in. Retrieved on October 12, 2012

Ministry of Tribal Affairs, Statistics Division, Government of India, (2013): Statistical Profile of Schedule Tribes in India

Mohanty PK 2004 *Enclopaedia of Primitive Tribes in India*, Delhi: Kalpaz Publication

Mohanty PK 2006 Encyclopaedia of Scheduled Tribes in India, Delhi: Isha Books

National Institute of Open Schooling: 2014 Annual Report 2010-11, *http://www.nios.ac.in*. Retrieved on October 15, 2014

National Institute of Science Communication and Information Resource, Council of Scientific and Industrial Research, New Delhi National Science Digital Library,: *http://www.niscair.res.in*. Retrieved on November 15, 2013

National Scheduled Tribes Finance and Development Corporation, Ministry of Tribal Affairs, Government of India, New Delhi, (2014): Twelfth Annual Report - 2012-13

Patnaik N 1982 Tribes and their Development: A Study of Two Tribal Blocks in Orissa, Hyderabad: National Institute of Community Development (NICD)

Planning Commission, Government of India, New Delhi: 2014 Empowering the Schedule Tribes for the Tenth Five Year Plan (2002-2007), *http://planningcommission.nic.in*. Retrieved on Decmber 11,2014

Planning Commission, Government of India, New Delhi: 2014 Five Year Plans and Annual Plans *http://planningcommission.nic.in*. Retrieved on December 11, 2014

Presler HH 1971 *Primitive Religions in India, Madras*: Indian Theological Library C.L.S.

Rajeeva 1989 *An Introduction to the Tribal Development in India*, Dehradun: International Book Distributors

Ramaiah P 1981 *Tribal Economy of India*, New Delhi: 13 Haratiya Adimajathi Sevak Sangh

Real and Radical Facts about India: 2014 Tribes in India: http://www.real-factsabout-india.com. Retrived on December 09, 2014

Remamani VS 1988 *Tribal Economy Problems and Prospects*, Allahabad: Chough Publications

Report Submitted by Society for Regional Research and Analysis, Gurgaon to Planning Commission, Government of India, New Delhi (2010): *Migration of Tribal Women: Its Socioeconomic Effects - An in-depth Study of Chhatisgarh, Jharkhand, M.P and Orissa*

Roy S 2006 Tribal Youth Dormitory: A Hiatus or Heuristic?, *Studies in Tribes and Tribals* 4 19–30

Schefler HW 1997 *Handbook of Social and Cultural Anthropology* (ed) JJ Honigman, New Delhi: Rawat Publications

Shah MA, Baviskar BS, and Ramaswami EA 1996 *Social Structure and Change*, New Delhi: Sage Publications

Sigmund F 1950 *Totem and Taboo: Some Points of Agreement between the Mental Lives of Savages and Neurotics*, New York: W.W. Norton & Company

Singh VRR, Mishra D, and Dhawan VK 2011 *Status of Joint Forest Management in India (as on June 2011)* Proceedings of National Workshop on JFM, 27-28 June 2011. Forest Research Institute. Indian Council of Forest Research and Education, Dehradun.

Srivastava RS 2005 Bonded Labour in India: Its Incidence and Pattern, Working Paper presented in International Labour Conference on Special Action Programme to Combat Forced Labour, Office, *International Labour Office*, Geneva.

Subramanya G 2003 *Social Problems with special reference to India*, Bangalore: Sapna Book House

Tiwari DN 1983 Development Strategy for Forests, Tribals and Environment, *Indian Forester*, 109 717

Vidyarthi LP 1972 Problems and Prospects of Tribal Development in India, *Indian Anthropologist*, 2 80–93

Vidyarthi LP 1986 *Applied Anthropology in India*, Allahabad: Kitab Mahal

Vidyarthi LP and Rai BK (eds) 1985 *Tribal Culture of India*, New Delhi: Concept Publishing Company.

Witchcraft Wikipedia; The Free Encyclopedia 2014 http://en.wikipedia.org. Retrived on January 20, 2015.